Conventional Label Printing Processes

Letterpress, lithography, flexography, screen, gravure and combination printing

Other Labels & Labeling books:

For the latest list please visit: **www.labelsandlabeling.com**

Conventional Label Printing Processes

Letterpress, lithography, flexography, screen, gravure and combination printing

John Morton
Robert Shimmin
4impression

Conventional Label Printing Processes
Letterpress, lithography, flexography, screen, gravure and combination printing

First edition published 2014 by:
Tarsus Exhibitions & Publishing Ltd

© 2014 Tarsus Exhibitions & Publishing Ltd

Printed by CreateSpace, an Amazon.com company.

ISBN 978-0-9547518-9-0

Contents

While every care has been taken to ensure the information, charts, diagrams and illustrations in this publication are correct at the time of publishing it is possible that technology, specifications, markets and applications, or terminology may change at any time, or that the editor's or contributor's research or interpretation may not be regarded as the latest accepted guidance in some parts of the world of labels.

The publishers therefore cannot accept responsibility for any errors of interpretation or for any actions, decisions or practices that readers may take based on the publication content and would advise that the latest industry supplier specifications, standards, legislative requirements, performance guidelines, practices and methodology should always be sought before any investment or implementation is made.

Foreword

It is the purpose of this book to provide the reader with a broad understanding of all the major conventional, mechanical printing processes used today in the manufacture of self-adhesive labels.

This book will explore the types of printing equipment and the principles of each of the main printing processes used. Conventional or mechanical printing refers to the printing process in which manually prepared screens / plates are used for printing.

Although conventional printing is under pressure from digital processes it remains a potent force in the sector and it is important that both end-users and those in the industry clearly understand where these processes fit within the sector and the part they play both now and in the future.

As the chapters unfold it will become apparent that each conventional printing process has its own characteristics and is able to deliver a range of benefits for particular applications.

It will also become clear that there are many hybrid manufacturing configurations on offer that can combine both conventional and digital processes to create unique products and print solutions.

The relative popularity of each printing process will always ebb and flow and some will become entrenched in their own particular niches. It is important however to have a fundamental appreciation of each process, the mechanics behind them and understand how they are evolving and changing.

It is a key aim of the book to provide this understanding and to demystify the complex terminology that can obstruct the learning process.

It should be noted that although digital printing will be referred to in this book, an in-depth analysis can be found in a separate book, Mike Fairley's *Digital Label and Package Printing.*

John Morton and Robert Shimmin
4impression Limited

About the Label Academy

This book is part of the recommended study material for the Label Academy, a global training and certification program for the label industry. The Label Academy was created by the team behind Labels & Labeling magazine and the Labelexpo series of events.

The Academy consists of a series of self-study modules, combining free access to relevant articles and videos with paid text books (both printed and electronic). Once a student has completed a module, there is an opportunity to take an online test and earn a certificate.

It is expected that a Label Academy qualification will become a standard in the industry – for printers/converters, suppliers, brand owners and designers – and assist in providing a benchmark. In addition to its own training, the Label Academy will aim to become a resource provider to the many existing educational programs in the industry. Accredited training courses will be promoted through the Label Academy website and books will be provided at discounted rates.

The Label Academy concept was pioneered by industry expert Mike Fairley. This was in response to a reduction in the number of dedicated printing colleges and the need to standardize training across the world. The label industry also has its own specific training needs – it has some of the widest range of materials, printing processes and finishing solutions of any printing sector.

We are also working with other training experts and authors to ensure that the Label Academy provides up-to-date and relevant training material for the industry.

The Label Academy is supported by the key trade associations, including FINAT, TLMI and the LMAI.

www.label-academy.com

Label Academy sponsors

Thank you to our founding sponsors, without whom this ambitious project would not have been possible:

Cerm
Cerm designs business automation software solutions to meet the specific demands of flexo and digital narrow web printers. Using the latest technology, our team's focus is on innovation and continuous improvement.

Our automation solutions support each step in the printer's integrated workflow – from estimating to production, shipment and data collection – and provide the feature and functionality printers need to gain efficiency and improve profitability.

Cerm inspires collaboration and helps printers remain competitive in the market and deliver the best products possible. We are proud to sponsor the Label Academy and contribute to the future of the narrow web printing industry.
www.cerm.net

Flint Group Narrow Web
Flint Group Narrow Web has the products, the solutions, and the technical experts to handle any print situation. Providing solutions for food packaging, sustainability, increased bottom line, efficiency, and uptime – delivering the basics needed to run a successful operation, and the expertise to go above and beyond to another level of success.

Our experts provide solutions to your printing problems with the innovative products and services that have made us an industry leader around the world. Wherever you are, we are – available to help you reach your business goals today and into the future.

Continuous improvement is paramount to Flint Group; we are proud to sponsor the Label Academy and the benefits it will bring to the future of our industry.
www.flintgrp.com

Gallus Group

The Gallus Group with its production sites in Switzerland and Germany is a leader in the development, production and sale of narrow-web, reel-fed presses designed for label manufacturers. The machine portfolio is augmented by a broad range of screen printing plates (Gallus Screeny), globally decentralized service operations, and a broad offering of printing accessories and replacement parts. The comprehensive portfolio also includes consulting services provided by label experts in all relevant printing and process engineering tasks. The Gallus Group is a member of the Heidelberg Group and employs around 430 people, of whom 253 are based in Switzerland. The group headquarters is in St.Gallen, Switzerland.

www.gallus-group.com

MPS Systems B.V.

Producing high-quality label printing depends on several factors; one of them is the operator of the press.

As a press machine builder since 1996, MPS Systems B.V. knows how important training and education on subjects like pre-press, label printing and finishing is. For label printers, it is critical that their operators keep up with pre-press and press developments in addition to label trends. Therefore, MPS sponsors the Label Academy, to advance operator's passion for printing, share expertise and help multiply benefits.

The MPS slogans of 'Printers First' and 'Technology with Respect' have always underlined the core philosophy of MPS from press design to operator satisfaction. We develop our presses with a strong focus on user-friendliness and respect for the press operator: Printers First.

www.mps4u.com

HP Indigo

HP Indigo is a global leader in digital printing, with a broad portfolio of digital presses and workflow solutions. Indigo's proprietary Liquid Electrophotography (LEP) technology delivers exceptional print quality for the widest variety of applications including labels, flexible packaging, shrink sleeves and folding cartons. HP Indigo's digital presses match gravure print quality satisfying the most demanding brands.

A division of HP Inc.'s Graphics Solutions Business, Indigo serves customers in more than 122 countries, including many of the top label and packaging converters worldwide.

www.hp.com/go/labelsandpackaging

UPM Raflatac

In a little more than three decades, UPM Raflatac has become one of the world's leading manufacturers of pressure sensitive label materials, developing and leveraging the latest innovations in adhesive technology. Our film and paper label stocks are used for product and information labeling across a wide range of end-uses – from pharmaceuticals and security to food and beverage applications.

We are an engineering driven company with industry-leading products known for their consistent high quality and top performance. We are also known for the high performing supply chain and undisputed leadership in the area of sustainability. UPM Raflatac's dedication to innovation, sustainability and top quality is matched only by our commitment to service excellence. We call it the Raflatouch.

www.upmraflatac.com

About the authors

4impression Training

4impression are specialist providers of training across a wide range of print and packaging related subjects. Staffed by industry trained tutors and supported by a network of print and packaging suppliers, the company delivers face to face to courses providing understanding of print processes, embellishments, materials, origination and finishing. Recently 4impression wrote the FINAT Educational Handbook which covers all aspects of self-adhesive label manufacture. They have also produced a comprehensive range of learning resources for the FINAT Knowledge Hub.

As authors of this book 4impression are uniquely positioned to offer additional personalised training to readers who require more insight into its content. The directors of 4impression, colleagues from their days working for the Jarvis Porter Group, are passionate about print and have a long track record in delivering courses to major packaging users and their supply chains.

John Morton

John has hands-on experience of all the major printing processes and has held operational and technical development roles at director level in the packaging sector. A qualified printer, John's career spans magazine production, commercial print, packaging and label production. Before joining 4impression John was actively involved in the Unilever advanced printing and decoration training courses attended by delegates from operations around the globe.

Robert Shimmin

Robert has held senior marketing and business development positions in the print, packaging and label sector spanning more than 20 years. He is a regular contributor of articles to the print and packaging trade press and has supported initiatives that seek to build awareness of the latest research and innovations emanating from UK universities. In addition to his involvement with 4impression he runs Shimmin Associates, a research and marketing consultancy offering support to both UK and international customers in the label and packaging sector.

Paul Jarvis

Paul Jarvis, formerly chairman of Jarvis Porter Group PLC, oversaw its growth to become one of Europe's leading packaging suppliers with a turnover in excess of £100 million, employing 1,600 people in 7 countries including the United States. Paul was a director and founder member of the Leeds Training and Enterprise Council and represented CCL Label on the main board of FINAT (the world-wide association for self-adhesive labels and related products). Paul provides strategic direction to the packaging and print sectors capitalising on his vast experience and global network of contacts.

www.4impression.com

Chapter 1

The changing landscape of label printing

Over the last thirty years there has been a startling evolution in both the equipment and printing methods used for the manufacture of self-adhesive labels.

In the early 1980s the predominant process used was semi-rotary and rotary letterpress. By the mid-1980s rotary letterpress made up almost 70% of all new narrow-web press installations each year. At this time flexographic process was deemed by many as inferior.

It was not long, however, before the use of flexography in the sector began to gain momentum, particularly in the USA and in mainland Europe.

Press manufacturers were producing well engineered equipment, but developments in inks, anilox rollers and origination were unable to keep pace.

Initially solvent based ink systems led the way, but it was the evolution in water-based and water-soluble inks and the use of photopolymer plates that enabled flexo to improve its quality and increase its share of the market. Towards the later part of the 20th century flexo had become the dominant process for new press installations.

Much of the early part of the 21st century has seen UV flexo rapidly rise to become the dominant technology for new press sales.

Today screen, litho and gravure printing continue to have specific niche roles to play in the label sector and their distinct advantages are often used in conjunction with the more mainstream flexo and letterpress processes on combination printing equipment. Offset litho continues to be the dominant process used in the manufacture of sheet and web-fed wet-glue labels, although gravure still plays a key role in web-fed production in some parts of Europe.

It is the recent growth in digital printing however, that has brought about the most dramatic changes to the label printing landscape and it is this process that is set to challenge conventional analog printing in developing economies. **(See Figure 1.1).**

KEY DEVELOPMENTS IN THE PRINTING SECTOR

A number of significant developments have influenced the changing fortunes of print processes used in the label sector.

Developments in die-cutting

The die-cutting of the label laminate to size and shape is a fundamental process involved in the manufacture of self-adhesive labels.

Most letterpress equipment originally relied on expensive flatbed die-cutting units as the chosen method of conversion. It was the flexo sector however, that took another leap forward with the introduction of cheaper and faster running rotary cutting units. Further progression was made in the 1990's with the development of flexible magnetic cutting dies manufactured using laser technology.

Flexible dies worked in conjunction with rotary magnetic cylinders to enable quick changeovers and

Trend in key European label press installations by technology (1980 to 2012)

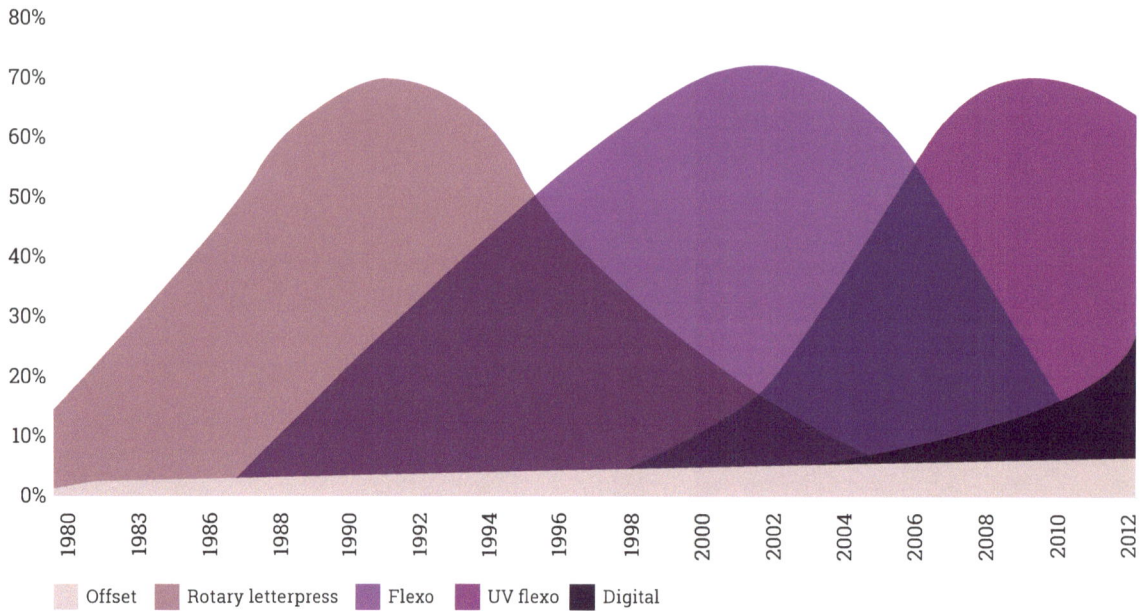

Figure 1.1 - Trend in key european label press installations 1980-2012 *(Source: Labels & Labeling)*

running speeds. **(See Figure 1.2).**

In the near future laser die-cutting developments (already widely used with the latest UV inkjet presses) are set to foster further advances in label printing.

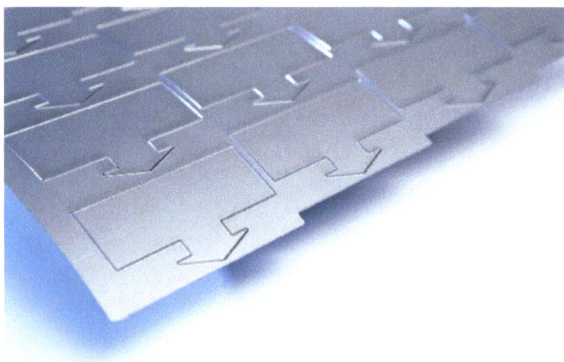

Figure 1.2 - Typical flexible die-cutting shim *Source: Wink*

Developments in digital repro

Rapid developments in digital reproduction and artwork have resulted in a quantum leap in the quality and speed of repro and plate making, which had positive impacts on all printing processes.

Digital image processing equipment made it possible for converters to produce their own artwork and stepped film efficiently in-house, which has resulted in improved control and output of jobs.

Developments in inks

A key development in the evolution of printing processes was the introduction of UV inks. Water-based inks proved unsuitable for keying to synthetic materials and it was the introduction of reliable UV curable inks that propelled flexo ahead of letterpress as the dominant process for self-adhesive label production.

The latest developments in LED curing however, offer interesting competition to other UV systems.

In 1993 however the E-Print 1000, the world's first digital color press was launched. In the mid-1990s reel-fed digital presses were introduced, with the Xeikon press launched in 1993. With the cost of entry falling, digital printing began to take off with three types of digital technology ... dry toner, liquid toner and inkjet at the fore. A number of factors boosted the attractiveness of the digital process including;

- **No origination costs**
- **Variable imaging and data**
- **Multi-versions and label variations produced economically**
- **Potential for personalisation**
- **Lower changeover times**
- **Reduction in set up and running waste**

Figure 1.3 - HP Indigo 30000 Digital Printing Press *Source: L&L Sep 2013*

GROWTH IN DIGITAL PRINT

Digital ink jet evolved in the 1970s and 1980s and was originally seen as a simple process for overprinting sequential numbering and simple variable data.

More than 2,200 digital label presses have now been installed worldwide (end 2013), since the first launches of this new technology in the mid-1990s.

Rising market share

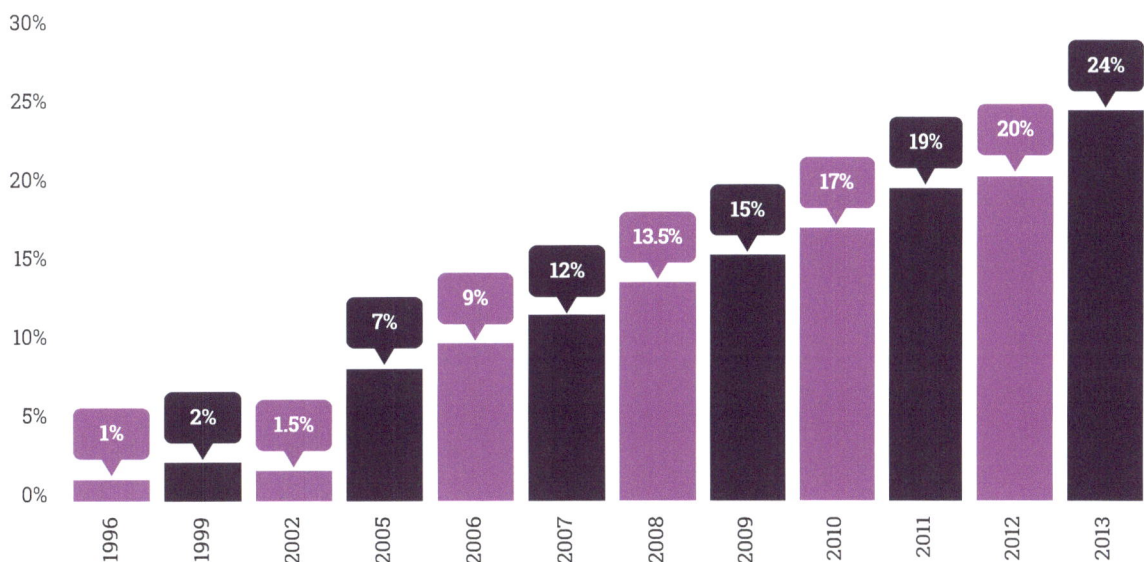

Figure 1.4 - How digital is impacting on the label industry *Source: Labels & Labeling*

Close to 350 new digital presses are now being installed each year into label printing companies in Europe, North and South America and Australasia. Today installations of new digital presses annually now make up some 24% of all narrow web label presses installed worldwide **(See Figure 1.4)**. Over the next five years annual installations are forecast to rise to at least 35% of all worldwide narrow web press installations.

Market share of new digital presses installed worldwide as a percentage of total new presses installed are stimated to be in excess of 35% in the UK and Europe by 2018.

Advances in ink-jet technology will undoubtedly see the digital process grow in stature and the launch of the nanographic* offset inkjet printing process is creating great interest in the sector.

* Nanographic - definition

Nanographic offset inkjet utilises nano-sized pigments to absorb much more light than other pigments, and thus permits images with ultra-sharp dots of extremely high uniformity and broad CMYK color gamut.

INTRODUCTION OF SERVO DRIVE MOTORS

The introduction of servo drive motors into conventional press engineering has resulted in higher quality, shorter set-up times and less waste. This has given conventional analog printing processes a timely boost against the rising tide of digital.

Servo drive technology for hybrid and single driven label presses may be axle driven with several gear ratios, axle driven by mechanical transmission elements or direct axle driven by assembled motor with no mechanical elements.

Servo drive technology enables manufacturers to produce presses with:

- **No main drive shaft**
- **No gears or gear boxes**
- **No mechanical main shaft**
- **No gears or bearer rings.**

Such presses offer fast set-up by recalling press data out of the job file, including length and cross cutting, register pre-setting, web tension pre-setting, web thickness setting, as well flying imprinting - that is text or image change without stopping the machine.

COMBINATION PRINTING

Combination printing is at the heart of innovation in the narrow web label sector, with the focus on adding-value to the label rather than cutting costs. It is particularly found in the wines and spirits, cosmetics and luxury goods market sectors.

Using a system of interchangeable cassettes, combination printing is the method of grouping a number of different printing and converting processes on one platform, in order to optimise the graphics and embellishments that can be produced in one pass **(see Figure 1.5)**. For example, the solid color coverage of screen with the halftone printing quality of letterpress, flexo or offset.

Figure 1.5 - A modern combination press *(Nilpeter)*

Combination printing focuses on the strengths and advantages of each printing method and converting process in order to maximise the visual appeal of the label and to enhance its graphic content. This factor alone reinforces the need for end-users, designers and converters to really understand the attributes of each printing process and how they can effectively be combined to deliver optimum results.

Today, hybrid combinations of different technologies within a single company are no longer the exception, but have become the norm.

LABEL PRINTING - THE FUTURE

A recent survey conducted by Labelexpo (2014) provided an interesting snapshot of those printing press technologies that were of greatest interest to visitors to their global exhibitions.

The survey highlighted flexo as the most sought after technology. **(See Figure 1.6).**

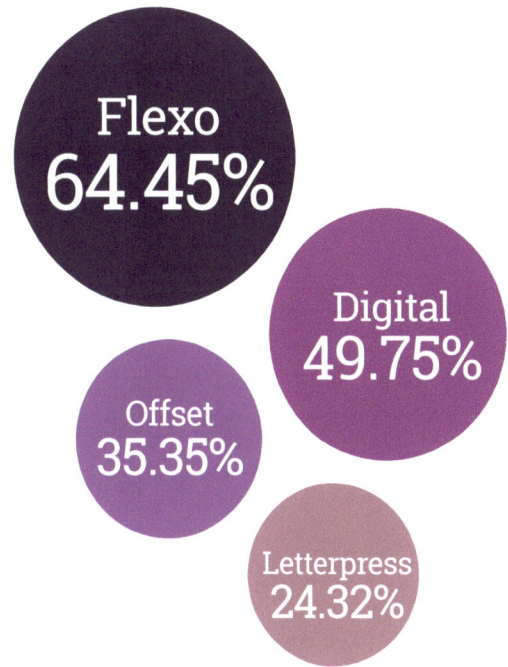

Figure 1.6 - Interest in printing technologies

Chapter 2

Label press configurations and ancillary equipment

This Chapter examines the main types of printing equipment used for the production of labels. In particular it will identify the different types of press configurations used in the sector, whilst highlighting some of their key benefits.

There is a wide range of ancillary equipment and processes that are fundamental to the successful production of labels and that deliver significant additional benefits.

The aim here is to provide a valuable insight into all of these whilst at the same time clearly explaining the terminology used.

TYPES OF PRINTING PRESS

There are a number of different press configurations used in the label industry. Most self-adhesive labels are supplied in the reel and are therefore produced on a reel-fed press. A smaller quantity of self-adhesive labels are manufactured in sheet form and these are produced either on a sheet-fed press or on a reel-fed press and then subsequently cut to the correct sheet size from the reel. **(See Figure 2.1).**

Reel supplied self-adhesive labels are printed, embellished and converted on a reel fed press (also known as a web-fed press) and produced as an in-line operation.

There are a number of different press configurations used in the manufacture of self-adhesive labels and this chapter explores some of the most popular ones.

Press configurations are determined by a number

of considerations:-

- **The type of printing processes required.**
- **The type of substrate to be printed and converted (paper or filmic)**
- **The range and weight of the substrate to be printed and converted**
- **The substrate characteristics**
- **The factory space requirement**
- **The capital cost**
- **The manning requirement**
- **Type of label product to be manufactured (multi-layer, printed adhesive etc.)**

The majority of web-fed label presses, regardless of configuration, would be equipped with all or some of the following pieces of ancillary equipment.

1. **In-feed reel stand or multiple stands - Flying splice unit**
2. **Web edge guidance unit**
3. **Web tension control**
4. **Corona discharge unit**
5. **Web cleaning unit**
6. **Static control unit**

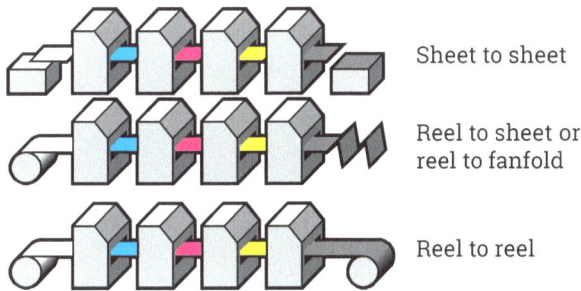

Sheet to sheet

Reel to sheet or
reel to fanfold

Reel to reel

Figure 2.1 - Typical sheet-fed and reel fed configurations used for label manufacture. *Source: 4impression*

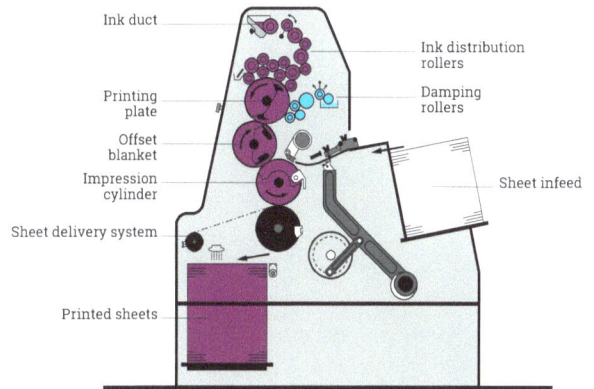

Figure 2.3 - Diagram of sheet-fed print unit.

7. **Print units**
8. **Inter-deck drying units**
9. **Die-cutting unit or units**
10. **Register control system**
11. **Sheeting unit**
12. **Matrix removal unit**
13. **Slitting unit**
14. **Outfeed rewind unit**
15. **Outfeed 'turret' auto rewind unit**
16. **Web turner bars**

Each of these ancillary items are explained in detail later in this chapter.

Figure 2.2 Shows a typical layout of a modern reel-fed press and the location and position of the ancillary equipment used (highlighted in purple).

CONFIGURATIONS USED TODAY

What are the different configurations of label printing presses?

The following pages illustrate some of the more common press configurations used today whilst highlighting the advantages and disadvantages of each type.

THE SHEET-FED PRESS

The typical sheet-fed press is configured with an auto sheet feeder unit called the 'feeder' and a sheet stacking unit called the 'delivery'. **(See Figure 2.3).** The print units are positioned between the 'feeder' and 'delivery'. Each print unit prints a single color and sheet-fed presses can be configured with the desired number of print heads as required. For instance a 6-color press will print up to 6 individual

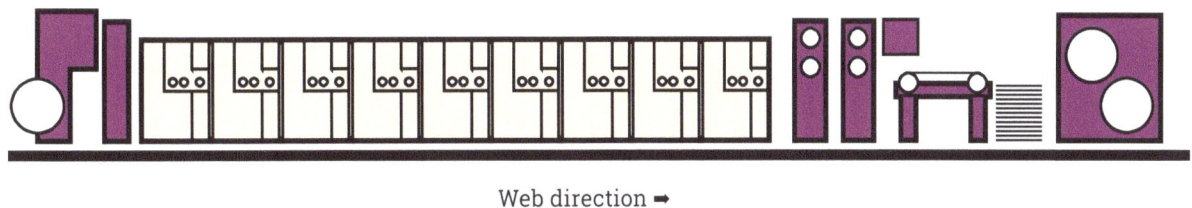

Web direction ➡

Figure 2.2 - Positioning of ancillary units on web-fed press

colors or less if required. The number of printing units can vary between one single unit, and in the commercial print market can be as many as 10 units.

Self-adhesive labels in sheet form are usually printed on multi-color presses and over-varnish is done either with an in-line varnish unit or alternatively on one of the printing units.

The vast majority of sheet-fed presses tend to use the litho process, but sheet-fed letterpress machines are still used today and are generally single color.

To aid output and manufacturing efficiency the modern sheet-fed press can be specified with an in-line coating unit which uses the flexo process to apply overall or spot coatings of varnish to the printed sheet. Other aids include automatic plate change, closed loop color and quality control systems and inter-deck drying which enables wet-on-dry printing between each print unit.

THE PLATEN PRESS

The illustration above **(Figure 2.4)** shows the Platen or Clamshell type of press configuration which is one of the earliest types of printing press. The press prints by the letterpress process, using either type or polymer plates. The inking rollers move up and down over the type or plate, which is located in a 'chase.'

The substrate, which is in sheet form, is fed onto the moving section of the platen (as indicated by the arrow on **Figure 2.4**). The location of the substrate within the press can be either manual or automated.

The back section of the platen (the bed) holds the 'forme' and remains static.

With the type 'inked' and the substrate in position, the platen comes together bringing the substrate into direct contact with the type thereby transferring the ink to the substrate. The platen then opens and the printed substrate is removed before the cycle begins again.

Because of the heavy construction and high impressional strength of this type of Platen configuration it is used extensively for the die-cutting of self-adhesive labels, in a sheet format.

THE SEMI – ROTARY 'STOP FEED PRESS

The semi-rotary print head differs from the full rotary

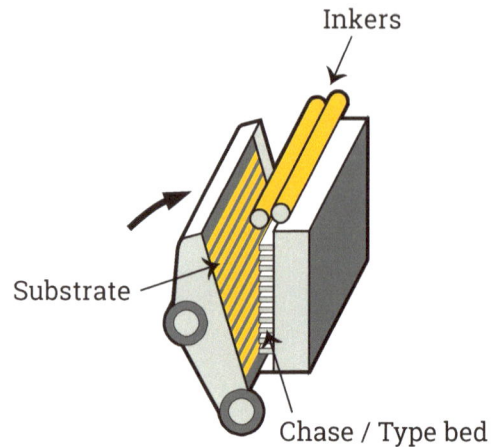

Figure 2.4 - Diagram of platen press (4impression)

press in two important areas. Firstly the printing head is positioned to print 'across' a 'stationary' web **(See Figure 2.5)** and not 'down' the web as is the case with a full rotary printing head. Secondly the semi-rotary head does not require print cylinders of differing diameter as does a full rotary head, which requires a print cylinder diameter that is compatible with the label size being printed.

The action of the web as it passes through the press is a stop – start movement with the printing cycle taking place when the web is stationary. As the web stops the print cylinder rolls across the web and transfers the image from the letterpress plate. The cylinder then rolls back to its start position and fresh ink is applied to the plate image before the web moves forward. This action is called the 'Pull' and the distance of the pull is set by the printing width of the plate cylinder.

This type of semi-rotary press configuration uses multiple letterpress printing heads and can also be fitted with flatbed hot foil stamping units, flatbed embossing and flatbed die-cutting units and waste rewind.

SEMI- ROTARY PRINTING

The key advantages and disadvantages of semi-rotary printing can be summarised as follows;

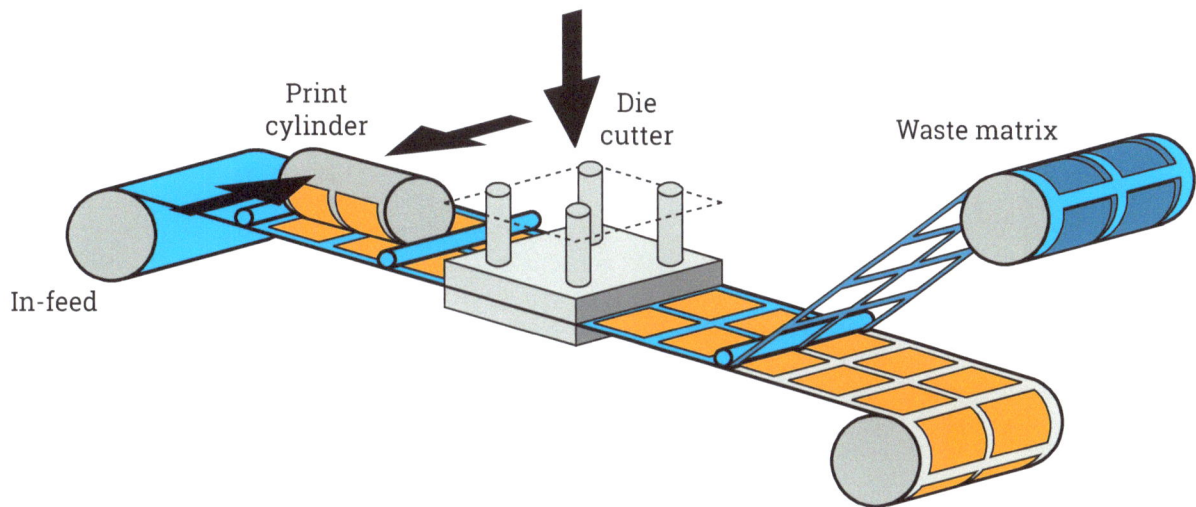

Figure 2.5 - Semi rotary stop feed label press

Advantages
- Does not require additional different diameter print cylinders
- Good quality print from the letterpress process
- Multiple embellishing and conversion facility
- Competitive preparation and tooling costs
- Good for short run work
- Easy press access for the operator

Disadvantages
- Slow running
- Letterpress process only

THE ROTARY 'INTERMITTENT FEED' PRESS

The Rotary Intermittent feed press also known as a 'translator' or 'reciprocating' feed, is a completely different configuration to the semi-rotary press. The press layout is the same as a full rotary web-fed press, but the difference lies with the web transportation system. With this system the web travels through the press with a forward and backward action and the printing cycle only takes place as the web travels forward. The print unit has a 'fixed' cylinder repeat length and always revolves in the same direction as the forward movement of the web. The print length is governed by the amount of web traveling through the press and this 'pull' distance is controlled by the translator feed system, which operates via servo driven cylinders and nip rollers positioned at the in-feed and outfeed sections of the press.

Because the print cylinders and/or offset blanket cylinders are not fully circular the gap in these cylinders allows the web to travel freely on the backward pull of the web and the fixed print cylinder repeat length removes the need for a stockholding of different cylinder sizes as required with the 'cassette' type litho, screen and flexographic processes.

The modern intermittent feed press uses the very latest printing technology and is used extensively for the manufacture of very high added value self-adhesive labels on a range of paper, film and metallic substrates. Offset litho is the print process most commonly used on this type of press, usually in combination with the screen and/or flexo process. These presses use high quality digitally driven servo drives and can be configured and retro fitted as required. Touchscreen operation with job data memory minimises make-ready times and waste on repeat jobs.

ROTARY INTERMITTENT FEED PRESS

The key advantages and disadvantages of the Rotary Intermittent feed press can be summarised as follows;

Advantages
- Can print in a variety of formats with minimal tooling/cylinder costs.
- Touchscreen operation with job data memory
- Quick setup times
- Low waste factor
- Only one operator required

Disadvantages
- High capital cost
- Slower running speeds

THE STACK PRESS

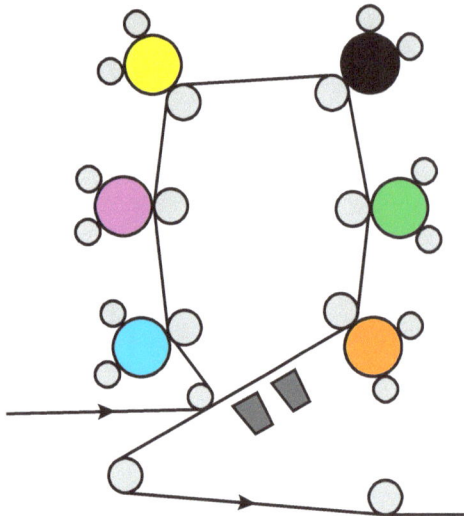

Figure 2.6 - Diagram of stack press. *Source: 4impression*

The term 'Stack' describes the way in which the print units are positioned onto the side frame of the press **(See Figure 2.6)**. These print units are 'stacked' on top of each other with sufficient space between each

unit to allow for an ink curing/drying unit to be located. The stack can be configured to provide twin stacks allowing 6 print units to be used.

When printing by the letterpress process each print head has its own set of inking rollers, plate cylinder and impression cylinder and because of the close proximity of the print units a short web path is achieved, reducing waste and print register errors. Other major advantages of the stack configuration are the reduction in factory space required and easy all round access for press operation and the internal engineering.

Figure 2.7 - Typical stack label press configuration. *Source: Edale*

The stack press configuration was instrumental in developing and increasing the manufacture of HAV (high added value) self-adhesive labels. This press configuration can produce high quality print on both paper and filmic substrates using combinations of the letterpress, flexographic and screen printing processes. The ability to produce filmic self-adhesive labels for the increasing 'clear on clear' label market and the ability to also rotary screen has given this press some considerable advantages over other types of label presses.

The 'stack press' can also be equipped with a hot foil stamping unit and flatbed and rotary die-cutting as an in-line operation. Its short web path and easy makeready make this type of press suitable for both short or long run work, with accurate registration quickly established and maintained throughout the run.

THE STACK PRESS

The key advantages and disadvantages of the 'stack' press can be summarised as follows;

Advantages.
- Small factory footprint
- Multi-process applications
- Good accessibility
- Easy operation
- Short web path – less waste
- Excellent web and registration control
- Single manning

Disadvantages.
- No offset litho capability
- Possible weakness with cantilever units when fitted with heavy tooling in die station.
-

THE CENTRAL IMPRESSION (CI) PRESS

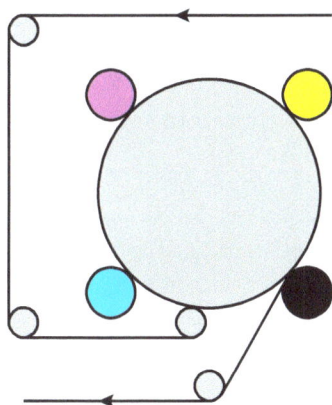

Figure 2.8 - Diagram of CI press. *Source: 4impression*

This type of label press features a single large diameter impression cylinder with each printing head located around it **(See Figure 2.8 & 2.9)** and an ink drying/curing unit positioned between each print

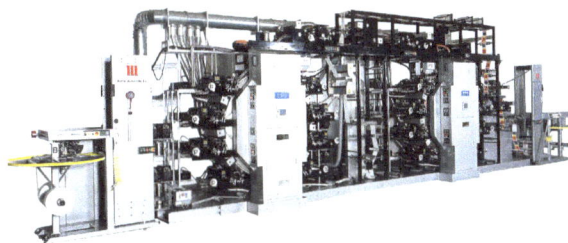

Figure 2.9 - Typical CI label press. *Source: KPG*

head. These presses can be configured with up to six print units.

One of the major advantages with the C.I. press is that the single large diameter impression cylinder is able to support the substrate throughout the printing of each successive color. The substrate remains in contact with the impression cylinder and as a result cannot move or distort, ensuring that excellent print to print register is maintained. This feature makes this press configuration excellent for the printing of very light filmic substrates.

THE CENTRAL IMPRESSION (CI) PRESS

The key advantages and disadvantages of the CI press can be summarised as follows;

Advantages
- Small factory footprint
- Excellent print to print registration particularly on very thin filmic substrates
- Easy operator access
- Hot air or UV drying
- Suitable for both the flexo and letterpress processes

Disadvantages.
- Potential for damage to the central impression cylinder
- Letterpress and Flexo only

COMBINATION – CI AND STACK PRESS

One of the hybrid presses which uses both Central Impression and Stack configurations is the Ko-Pack label press. This type of press configuration is very versatile and used for the manufacturing of complex multi-layer labels, pouches, coupons and booklets

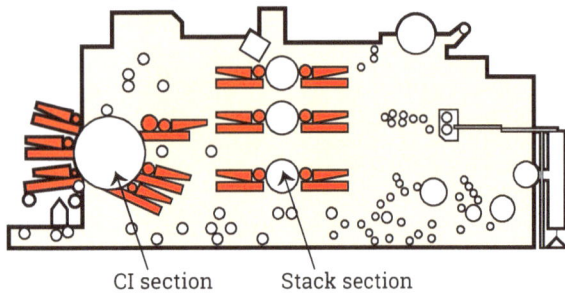

Figure 2.10 - Diagram of combination CI and stack label press

that may require the printing and converting of multiple webs, multiple folding, hot melt gluing, over-lamination and multiple die-cuts.

A typical press specification for the dual configuration shown in **Figure 2.10** would be :-

2 unwinds stands, web cleaning with corona discharge, 12 printing positions comprising of 11 letterpress units, 3 flexo units plus 1 rotary screen unit, interdeck UV drying between all the print units, 3 plow folding units, hot melt gluing unit, over-lamination system, and finally 3 rotary die-cutting units, matrix rewind, multiple slitting, sheeting and conveyor unit and product rewind stand.

THE WEB-FED MODULAR LABEL PRESS

The modular web-fed press prints onto a continuous running web and not as a single sheet. The unprinted reel is placed on the in-feed section of the press (this is called the 'unwind stand') and the printed and converted reel exits on the rewind section of the press, called the 'rewind stand'. The web passes through each individual printing unit and any print register adjustment that is required can be carried out from a central console or at each individual print station. These adjustments can be done manually or via the press auto registration system. **Figure 2.11** shows a 4-unit web-fed modular configuration, printing in the process colors of CMYK, cyan in unit 1 – yellow in unit 2 – magenta in unit 3 and black in unit 4.

The printing head is positioned on the upper part of each unit and the die-cutting and embellishing stations are positioned between the last printing unit and the rewind unit. In addition a drying/curing unit is positioned at the base of each print unit. This is called an inter-deck drying system which allows each individual color to be dry before the next color is printed. This is called wet on dry printing.

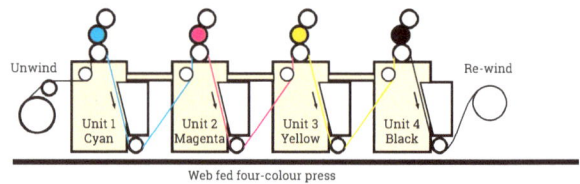

Figure 2.11 - Web-fed full rotary press. *Source: 4impression*

This modular type of press configuration is very versatile in a number of ways. Extra print, embellishing and conversion units can be retrofitted if required and the press layout gives the operator easy access for job changeovers and makeready

A standard specification for this type of press would include an unwind unit, web tension control, web cleaner, corona discharge and manual or auto print register, die-cutting unit, matrix rewind and outfeed rewind unit.

The modular press can be operated as a dedicated 'single' print process machine or as a multi-process platform capable of total process interchangeability. However it should be noted that the majority of these modular label presses are dedicated UV flexo.

THE WEB-FED MODULAR LABEL PRESS

The key advantages and disadvantages of the web-fed modular label press can be summarised as follows;

Advantages

- Easy access for operator
- No multi-color restrictions
- No multi-process restrictions
- Suitable for easy retro-fitting of ancillary equipment
- Single manning

Disadvantages

- High capital cost
- Greater factory space requirement
- Requires sophisticated web and registration controls

THE MODULAR MULTI-PROCESS 'COMBINATION' PLATFORM PRESS

One type of press configuration in particular has revolutionised the printing of self-adhesive labels. The modular multi-process 'Combination' platform press offers both end users and converters a range of significant benefits.

A full chapter has been devoted to Combination printing (see Chapter 8) but a brief description of this press type is provided below.

The 'printing' processes that are compatible with combination printing and can be used in any combination are:-

- Offset lithography
- Flexo
- Rotary screen
- Gravure
- Letterpress
- Digital

The 'embellishing and converting' processes that are compatible with combination work are:-

- Hot foil stamping
- Cold foiling
- Embossing
- De-lamination and re-lamination
- Die-cutting

Figure 2.12 illustrates a typical layout of a modern combination press. The diagram shows a comprehensive range of print processes, embellishments and conversion equipment and the

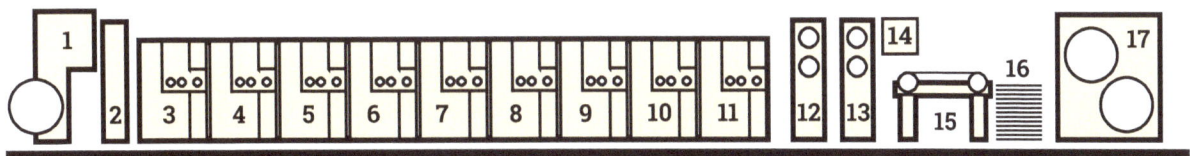

Figure 2.12 - Diagram showing position of units within combination press

Combination printing allows a wide variety of print processes to be utilised on a single press. Each of the various print technologies has its own advantages and disadvantages and combination printing focuses on the strengths and advantages of each printing process.

Combination printing is the method of utilising the different printing, embellishing and converting processes on a single platform, giving full interchangeability of these processes in order to optimise the graphic advantages offered by each process.

location of each unit.

Combination presses do not necessarily require all of the processes named in this section.

1. Automatic unwind and splicing unit
2. In-feed tension control
3-11 Multi-process stations for cassette system for Offset litho – Flexo – Screen – Gravure - Digital

12. Rotary die-cutting or embossing unit

13. Rotary die-cutting unit
14. Waste matrix removal
15. Sheeter unit
16. Sheet stacker unit
17. Rewind unit single or auto turret rewind

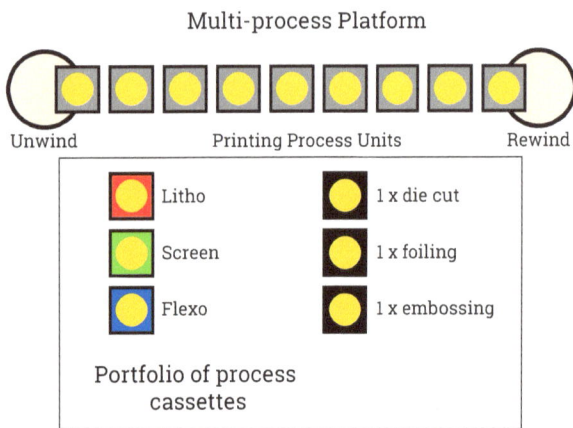

Multi-process Platform

Unwind   Printing Process Units   Rewind

Litho   1 x die cut
Screen   1 x foiling
Flexo   1 x embossing

Portfolio of process cassettes

Figure 2.13 - Ilustration showing base press with portfolio of interchangeable units. *Source: 4impression*

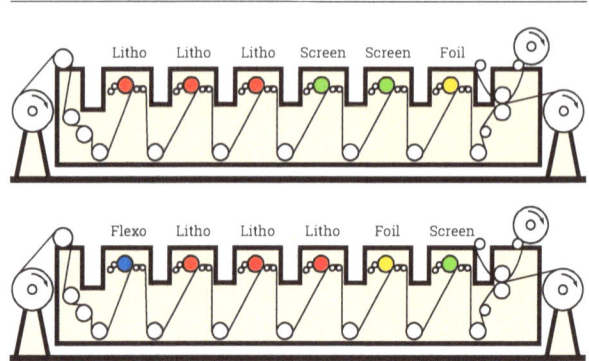

Figure 2.14 - Illustration showing typical unit combinations. *Source: 4impression*

Figure 2.13 shows a typical cassette specification that would be used to give the printer a range of graphic combinations.

The combination or platform press uses a system of interchangeable cassettes. The system allows a quick and uncomplicated method of providing the printer with a number of graphic combinations which can be easily assembled on the press allowing full interchangeability between the processes

The pneumatics, electrical and electronic systems used on a platform press allow easy changeover of each cassette so that all the printing, embellishments and conversion processes can be carried out as a one pass in-line operation. Graphic embellishments such as embossing, hot foil stamping and lamination are also in cassette format.

Figure 2.14 illustrates the flexibility offered by a combination press. The two illustrations show the same press in different cassette configurations giving the facility to use the most suitable printing and processing method.

The configuration of the combination press is identical to the modular web-fed label press and the use of servo driven motor technology and sophisticated digital control systems is now used extensively by the manufacturers of both the modular label press and the multi-process combination press. This kind of technology gives the operator the facility to use the computerised database to save and reuse the data that has been applied to a specific job, for example, previous press settings, ink control settings, web tensions, anilox roll specification, press speed etc.

MODULAR MULTI-PROCESS 'COMBINATION' PLATFORM PRESS

The key advantages and disadvantages of the modular multi-process 'Combination' platform press can be summarised as follows;

Advantages
- Ability to produce very high added value labels
- Process flexibility
- New and unique graphics
- New market opportunities

Disadvantages
- Higher investment levels for the press and related equipment
- Higher pre-press complexity and costs
- Higher skill level requirement, knowledge of different print processes requiring additional

training

- Possibility of additional staffing to support combination printing
- Requirement for knowledge of different ink technologies
- Training of sales staff to understand the benefits of combination presses and the potential to sell added-value
- Limitation on press capacity and job flexibility with a single combination press.

ANCILLARY EQUIPMENT

Whatever printing press or configuration is used there is a wide range of ancillary equipment that can be utilized, some optional and others more fundamental to the efficient running of the press. Although reference to some of this equipment will have already been made earlier in this chapter the following section provides a useful overview of the ancillary equipment used in the printing of reel-fed labels and will highlight their benefits.

1 IN-FEED REEL STAND

The in-feed reel stand holds the unprinted reel and has the facility for manually joining the end of the expiring reel to the end of the new reel. This operation is called 'web splicing'. It is essential that this spliced joint is positioned at a suitable angle to ensure that the converted web and waste matrix does not break at the point of removal. Single sided adhesive film tape is used for the splice.

2 FLYING SPLICE IN-FEED UNIT

The Flying splice unit **(See Figure 2.15)** allows the automatic joining together of two webs. Splicing is the procedure of attaching the leading and trailing edges of the web in the cross direction by the use of an adhesive strip. The cross cut is usually positioned at an angle to reduce the possibility of a web break whilst the press is running.

A flying splice facilitates the automatic joining together of two webs while the web is in motion and is most commonly used when reels are changed from an expiring roll to a new roll. This is done without stopping the machine, either at the unwind or the rewind ends of a web-fed press.

The key benefit of the flying splice unit is that it allows the press to continue running.

Figure 2.15 - Flying splice unit. *Source: Martin Automatic*

3 WEB EDGE GUIDANCE UNIT

A device on a web-fed label press that keeps the web traveling straight and true as it passes through the machine. Web guides **(See Figure 2.16)** play an

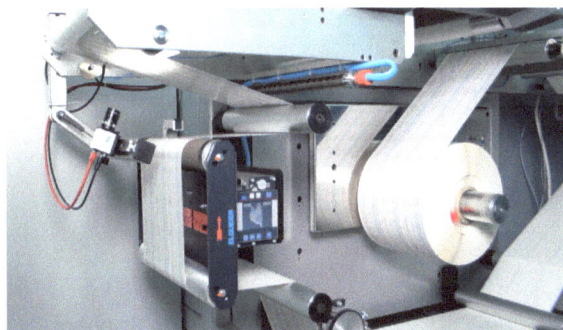

Figure 2.16 - Web edge guidance. *Source: Erhardt+Leimer*

important part in the control of the web through a press and may consist of photo-electric, UV light or ultrasonic sensors that are focused on the edge of the web. These sensors feed data back to either an adjustable pivoting roller, which guides the web to the left or right as required, or to the in-feed reel stand which can also move to the left or right.

Reaction speeds from the system should match the speed of the press to avoid any over-correction. If this is not synchronized then the web may move erratically from side to side thereby creating problems during printing and finishing operations.

4 WEB TENSION CONTROL

Web tension is the amount of pull or tension applied to the web as it passes through the press. Poor tension control will result in registration problems in the printing, embellishing and converting processes as a result of side to side web movement, substrate stretch or creasing.

During the unwinding of a roll on a converting or web-fed press the paper, film, foil or laminate material being used is under tension or stress as a mechanical force operates to extend, stretch or pull the web apart. This tension needs to be controlled. Too little tension makes register control more difficult; too much tension may lead to the material stretching or growing in length.

Color to color registration, size of label, position of any perforation or punching, etc. will all be affected by inadequate tension control; the more flexible the material the more sophisticated tension control becomes essential.

The degree of control that should be built into a press will involve an additional investment at the time of purchase, but should be more than repaid through increased productivity throughout the life of the press.

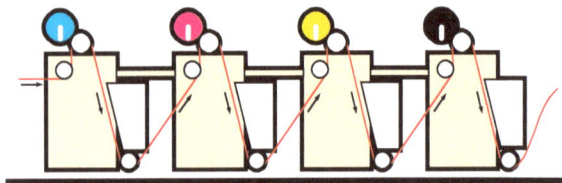

Figure 2.17 - Press with correct web tension and register aligned (4impression)

A typical tension control system on a press unwind incorporates sensing rollers mounted in a pivoting floating bracket balanced against the tension setting. The bracket and rollers change the floating position by pivoting in response to changes in factors such as roll diameter, running speed, acceleration of the web and friction changes within the braking system. The pivoting motion of the bracket transmits information that constantly adjusts the brake force to keep the tension in equilibrium.

Figure 2.17 shows a web-fed press running with the correct web tension and the print cylinders in 'register' on each printing cycle.

Figure 2.18 shows a press running with an incorrect web tension setting and shows the print registration 'hunting' to establish the correct print register. This problem, if unresolved can lead to considerable amounts of waste.

Thinner filmic materials require a very sensitive web tension control. The use of servo technology enables good control of the web and the facility to run the press at faster speeds without compromising the web tension.

Servo-driven controls make web control much easier for the press operator and eliminates the need for making manual changes to the web tension during the print run.

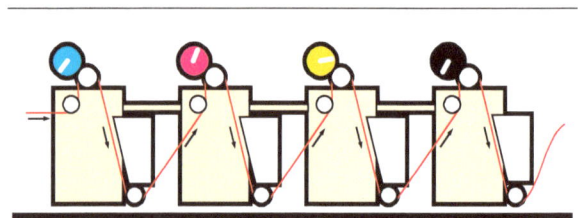

Figure 2.18 - Press with incorrect web tension and register misaligned (4impression)

Tension sensors monitor the changing weight and torque of the in-feed reel and the diameters of the rewind roll to maintain the right web tension. Transducer rollers automatically calculate these parameters and the unwind reel braking system will automatically adjust to prevent 'web wander'. Some of the common causes for web tension problems are;

- The general condition of the press
- Changes in roll diameter during the print run
- Elliptical shaped substrate reel

- Badly wound roll
- Slipping of the inner core
- Variations in the substrate caliper (thickness) within the reel. This occurs when there is variation between the left and right halves of the reel

Good tension control
- Improves print quality
- Improves web tracking
- Improves lamination quality
- Minimizes web breaks
- Allows thinner film stocks to be run

Bad tension control
- Causes bad register
- Allows web weave
- Causes product curl
- Creates downtime

5 CORONA DISCHARGE

In corona treatment, the surface of film or other material to be treated is bombarded with electrons.

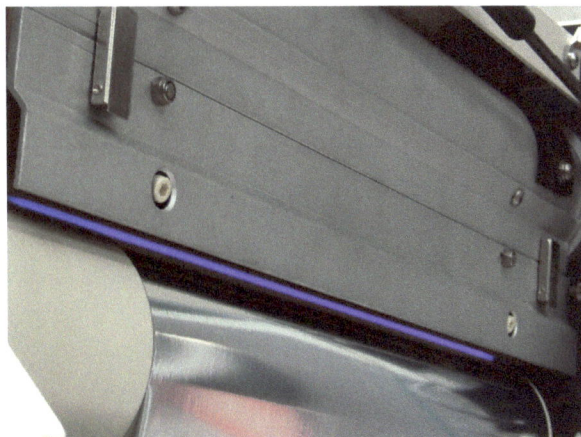

Figure 2.19 - Corona discharge unit *Source: Enercon*

These leave the electrode source and are accelerated under high tension towards the passing web material. In doing this they collide with air molecules which transmit light and react in part to ozone and nitrogen oxide. When the electrons come into contact with, say, a polyethylene film, they have so much energy that they can break the bond between the carbon hydrogen or carbon-carbon. Reactions with the corona gas take place at these free radicals, mainly towards oxidation, with the polar functional groups thus formed providing the basis for adhesion of applied printing inks, adhesives, lacquers, etc.

Ease of operation, ease of maintenance, operational reliability and simple handling are the key criteria for selecting corona equipment.

Equipment should also be easy to install within a short period of time and must include practical functions and control features.

Corona Discharge units are usually positioned prior to the point where the web enters the first print unit and close to the web cleaning unit. The purpose of the corona surface treatment is to increase the surface wettability of the substrate (generally filmic) through an electrical discharge. The low surface energy of polymer-based substrates often leads to poor adhesion of inks, glues and coatings. To obtain good adhesion it is necessary to increase the surface energy of the substrate.

Depending on the requirement, one or two rollers can be installed above or below the web for a double-sided pre-treatment. **(See Figure 2.19).**

6 WEB CLEANING UNITS

When selecting a web cleaner, whether contact or non-contact, one of the crucial factors is to make sure that the cleaner breaks the layer of air that is held by the moving web and which can hold contamination on the substrate surface.

Contamination can come from paper dust, factory dust, airborne contamination etc and will be attracted to the surface of the substrate and in turn will be picked up by the inking rollers and also the surface of the image plate causing print defects.

To break this contamination layer, different systems are used including powerful airflows, air turbulence or actual contact with the web.

Some web cleaning systems use a tacky roller, which is in contact with the web. Any debris that is picked up by this roller is transferred to a second, more adhesive roller.

The benefit of this technology is that it is relatively easy to install and gives good results.

7 ANTI-STATIC CONTROL UNITS

Figure 2.20 - Anti-static unit

Static electricity is the result of an imbalance between negative and positive charges on the surface of the substrate, and can be a problem when processing plastics substrates.

The static charge builds up on the surface and needs to be discharged by a combination of an electrical charge and ionized air.

Anti-Static units **(See Figure 2.20)** are used to remove the static charge that, in the case of web-fed presses, is generated by the movement of the web as it comes into contact with the web path rollers.

This issue can be particularly troublesome with filmic substrates with static electricity generated whenever two surfaces are in contact and are then separated, thus causing a dragging or holding effect. This problem can also affect sheet-fed applications. Static causes the sheets to be electrically attracted because of the opposing charges on each sheet. One sheet has a positive charge and the facing side has a negative. The effects of static electricity can be felt and heard and will often be seen as a large spark as the static is discharge or earthed.

8 INTER-DECK DRYING UNITS

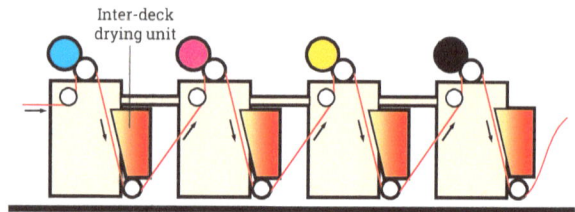

Figure 2.21 - Location of inter-deck drying unit. *Source: 4impression*

The modern self-adhesive press is fitted with inter-deck drying heads allowing each individually printed color to be cured/dried before printing the next color. This is called printing wet-on-dry and eliminates the problems which can be encountered when printing wet ink onto a wet ink.

The drying or curing head is positioned so that the printed substrate is exposed to the drying head for the correct dwell time to ensure that the ink is fully cured/dried **(See Figure 2.21)**.

There are three types of drying systems used in the self-adhesive label industry: Infra-Red, which uses IR lamps to create heat; Hot Air, usually created by IR but with the addition of blown and extracted air; and UV, that uses ultra violet lamps to cure inks formulated to cure/dry when exposed to ultra violet light.

9 DIE-CUTTING UNIT

The die-cutting unit is designed to hold a rotary die-cutting tool **(See Figure 2.22)** in a stable position, at the required pressure to perform the cutting action. This unit has to be of substantial construction and be capable of applying a consistent and accurate pressure throughout the job run.

The facility to make very fine adjustment is very important and modern units are fitted with pressure measurement equipment **(See Figure 2.23)** which allows the operator to monitor the pressure being applied to the die and avoid any overheating of the die 'bearers' and subsequent damage to the die. It is important that the die bearers are adequately lubricated throughout the print run.

located on the bottom platen. The substrate moves between the two platens using a reciprocating stop - start action; when the web is in the stationary position the platens come together under pressure and the cutting action takes place.

10 REGISTER CONTROL SYSTEM

The registration control system on the press is responsible for maintaining the accurate positioning of each color being printed. The system has to be capable of dot to dot accuracy between each print unit, print to embellishing and print to die-cutting / conversion, throughout the print run.

Register may be manually or electronically controlled and the system has to be able to allow four adjustments of the print cylinders/unit whilst the press is running, i.e. side movement left – right and circumferential advance and retard.

Manual systems are totally dependent on the operator to make all the necessary adjustments, but with automatic register control, the system will read the position on each printed color and automatically make any adjustments that may be required.

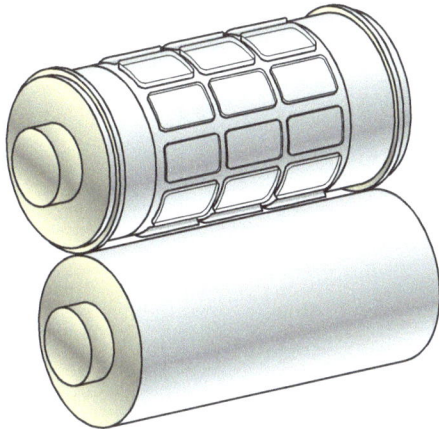

Figure 2.22 - Rotary die and anvil roller

Self-adhesive label presses can also be fitted with a flatbed die-cutting unit. This type of profile cutting unit uses a less expensive ruled die tool. The flatbed

11 MATRIX REMOVAL UNIT

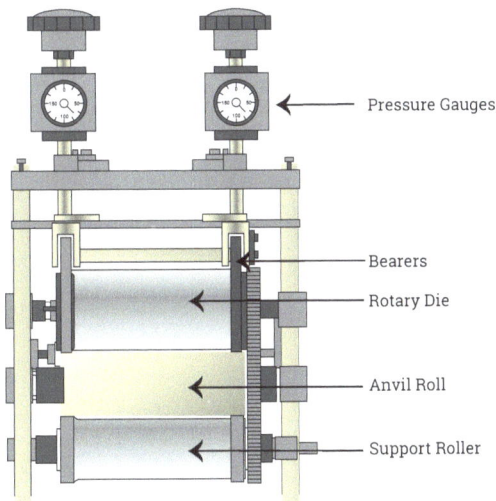

Figure 2.24 - Matrix removal unit

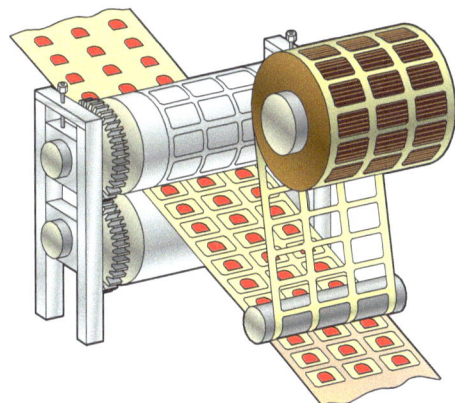

Figure 2.23 - Diagram of die-cut unit

unit works on a platen principle with the cutting die located in the top platen and the cutting plate

The removal of the waste matrix can be carried out as an in-line operation on the press or as an off-line operation.

There are two methods of matrix waste collection, it can be rewound onto a waste winder system as in **Figure 2.24** or extracted through a vacuum waste unit.

Some label shapes can prove very difficult during the matrix removal process and careful consideration needs to be given to the job layout to ensure that the removal of the matrix is trouble free and allows consistent running speeds. A number of presses are equipped with a driven roller placed just above the matrix take off point. The pull speed of this roller can be adjusted to increase or decrease the tension applied to the matrix allowing the operator more control of the matrix removal process.

12 SHEETING UNIT

Sheeting units are located at the unwind end of the press and allow the converting of printed substrates from a reel into individual sheets, as an in-line operation.

The material, in reel form, is fed from the unwind, passes through the printing and converting stations into a sheeting station which then sheets-off the printed material into specified lengths. The cutting operation can be done with a dedicated sheeting head which is fitted with a rotating cutting blade which passes over a fixed 'bottom' blade to make the cut.

The speed of this type of cutting head can be adjusted, which allows differing lengths of sheet size to be cut.

Cutting can also be done using a rotary die fitted with a cutting blade. This type of sheeting tool fits into the existing die-cutting unit and runs at a fixed speed. The cut is made when the cutting blade comes into contact with the anvil roller which is an integral part of the die-cutting unit.

13 SLITTING UNIT

The function of the slitting operation is to remove the unwanted edge of the web or divide the web in the lengthwise direction, to produce two or more narrower webs. The printed reels coming off the press are then slit into even narrower reels for automatic application on to the product.

Figure 2.25- Slitting Unit. *Source: ABG*

The slitting unit is generally positioned prior to the press rewind unit, but can also be positioned at the unwind end of a press to ensure accurate web widths enter the press. The slitting unit will be adjacent to a waste extraction unit which is used for the removal of unrequired web edge waste. The cutting heads can be of a rotary or razor cut construction.

There are two types of rotary cutting actions, the crush cutter and the scissor cut. The rotary crush blade cuts against a bottom anvil roller or segment, whereas the scissor cutting head has two rotary blades that are in contact with each other and rotate to create the scissor action **(See Figure 2.25)**. Razor units allow the fixing of sharp razor type blades which cut the substrate as it passes through the unit.

Modern slitting units are now capable of automatically positioning the top and bottom cutters, working from a digital program. These units eliminate the need for manual setting of the slitters.

14 REWIND UNIT

The outfeed rewind unit is placed at the end of the press and is usually the last process on the press. This unit can be a single shaft rewind or a twin shaft rewind and the amount of rewind torque/tension can be adjusted during the job run.

The facility to vary the torque is a very important factor particularly if heavy or lightweight stocks are being converted.

Reels can be rewound too tight or too slack and this problem can cause adhesive bleed and substrate creasing or even cause the reel to collapse whilst the press is running.

The shafts on the rewind unit can be rotated whilst the press is running allowing manual or fully automated splices to be done and thereby eliminating the need to stop the press during the job run.

15 TURRET AUTO REWIND UNIT

The auto rewind unit known as a 'turret rewind' can be used as an off-line or in-line facility. The purpose of a turret rewind is to produce self-adhesive labels in a finished format to a specified reel size or quantity of labels, ready for dispatch to the customer. The strips of labels which are created after the slitting process are rewound onto individual cores. As the reels reach the specified diameter the turret revolves to a point suitable for automatic splicing to take place. As the reel reaches the correct diameter or quantity of labels the new cores, which are adhesive coated, start rotating and a striker blade moves down and cuts through the ribbons, pressing the leading edge of the ribbons onto the new set of cores. If the rewind unit is positioned 'in-line' on the press there is no secondary process i.e. slitting.

Q.C. checks at the rewinding stage complete the operation and the finished labels, on the correct core size, are ready for product application.

The process is just the same when done off-line.

16 TURNER BARS

The function of the turner bars is to change the direction of a moving web to allow the printing of both sides of the substrate, the web turner can also change the direction of the web to allow a side exit from the press so the web can be diverted into a secondary conversion unit.

Figure 2.26 – 2.27 shows a web traveling from the right and passing through a series of angled rollers. Some of these rollers rotate and some are static. The static rollers are air assisted allowing the web to ride on a cushion of air. The web then exits on the left exposing the reverse side of the web. This facility helps change the direction of the web and allows printing on both sides of the substrate to take place.

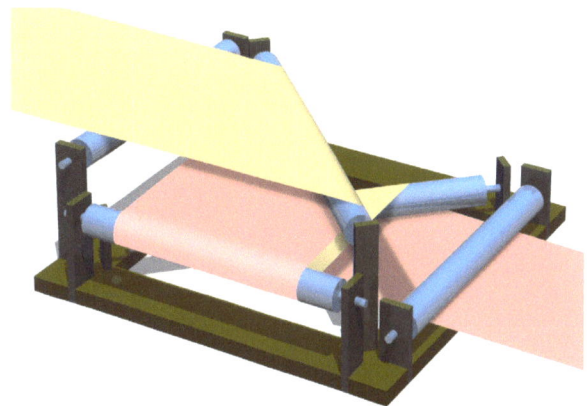

Figure 2.26 - 2.27 - Turner bar units. *Source: MPS*

Chapter 3

The letterpress printing process

The letterpress printing process takes its name from the method by which the process was used, i.e. 'letter - pressing'. Johannes Gutenberg is credited with the invention (circa 1440), but as early as the 2nd century the Chinese and Japanese used the 'letterpress' method to print using engraved wooden, porcelain and metal blocks.

The major development in letterpress printing was the use of woodcut and cast movable and reusable letters, known as 'type'. The letters were handset into lines to form pages and the pages locked into a 'chase' to create the 'forme' ready for fitting into the printing press.

The mechanisation of letterpress printing began with wooden printing presses based on a platen type configuration with the 'forme' being laid on the bottom platen. Ink was hand applied to the surface of the type with a leather covered ink ball and paper was then carefully laid on top of the inked type before pressure was applied by the top platen using a large threaded screw. This revolutionised printing as it allowed the mass production of books and documents. One of the most famous printed books was the 'Gutenberg Bible,' of which the first 200 copies were printed in 1455.

By the very end of the 18th century steel-made presses were being developed that used a knuckle and lever arrangement instead of the screw and the manufacture and use of inking rollers speeded up the process, paving the way for further automation.

MARKETS
The use of letterpress in the commercial print market is now small and shows a declining market usage, however letterpress is still of importance in the self-adhesive label sector. The label industry trend towards shorter runs has created a market for intermittent feed letterpress technology on shorter run jobs which require complex in-line multi-process printing, embellishing and conversion.

Typical applications have included short runs of high added value wine, spirit and cosmetics labels.

A lower capital investment makes it a popular process in the developing countries and whilst the quality of letterpress printing has never been in question, over the last two decades it has been displaced by other printing techniques. Preparation time for letterpress can be quite lengthy and its use is being overtaken with faster and more efficient printing methods such as the offset lithographic and flexographic processes.

In the more mature European labels markets, surveys show UV flexography has overtaken letterpress as the leading narrow web process in the global narrow web market. But it would be an error to think that this is an obsolete technology and there is still a large base of both rotary and intermittent fed letterpress machines running label work on a daily basis. The development of semi-rotary, servo-driven systems makes the process very competitive on short runs that require complex embellishing and converting, particularly when matched with flatbed die-cutting. Converters are often reluctant to make the shift from letterpress to flexography because of the need to change technologies and the more expensive ink and plate systems associated with other printing methods.

PRINCIPLES OF PROCESS

Printing is by a relief image plate. The relief (print) areas receive a film of high viscosity ink and under a controlled pressure known as the 'impression,' the image is transferred direct from the plate to the substrate. **(See Figure 3.1).**

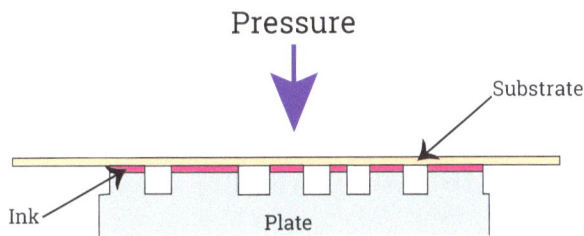

Figure 3.1 - Principles of process

Figure 3.2 shows the layout of a rotary letterpress unit of the type used in the label industry. At the top of the unit you can see the ink reservoir known as the ink duct, which distributes a controlled amount of ink to the inking distribution rollers. A consistent film of ink is applied to the printing plate by the forme rollers/ink train.

The substrate being printed travels between the rotating plate cylinder and the impression cylinder and the printing pressure is adjusted by increasing or decreasing the plate cylinder pressure to the substrate. There is very little 'give' in the printing plate, so in order to achieve a good ink film transfer and a sharp image, the pressure between the printing plates and the impression cylinder or flatbed requires a very careful setting. A skilled operator will vary the hardness of the impression roll ensuring that the plate to substrate contact is a 'kiss' touch. Careful balance of pressure between plate and impression roll is crucial as it determines the print quality. Too much

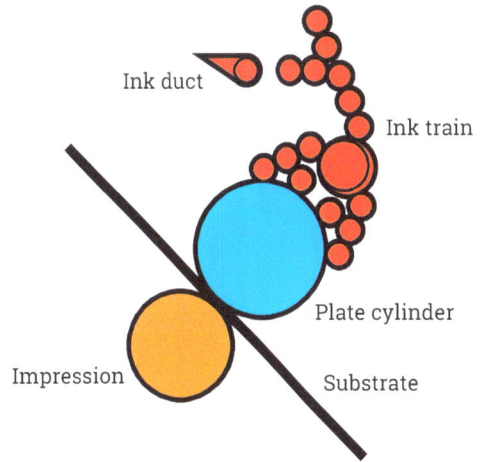

Figure 3.2 - Layout of rotary letterpress unit

impression creates a squashed or halo effect and too little impression creates missing dots and poor print.

Accurate plate inking roller settings are very important and impression adjustments should be made throughout the press run to make sure the correct printing pressure is maintained.

Letterpress printing uses a paste ink held in an ink duct and this is transferred to the surface of the printing plate through a series of ink distribution rollers. These distribution rollers rotate and oscillate from side to side, breaking down the ink film and ensuring that correct ink film weight ink is deposited on the printing plates with each revolution of the press.

LABEL PRESS CONFIGURATIONS USED FOR LETTERPRESS PRINTING

Figure 3.3 - Principles of process

Letterpress label printing presses come in a number of different configurations - platen, flatbed, semi-rotary and full rotary. All of these press configurations are used today for the manufacture of self-adhesive labels **(See Figure 3.3)**.

LETTERPRESS - PLATEN-TYPE
A platen press is made up of two flat surfaces called the top bed and the bottom bed **(See Figure 3.4).** The substrate to be printed is positioned on the top bed and the forme/plate is locked onto the bottom bed. The inking rollers traverse down the face of the plate and the plate or type is inked. The top and bottom bed come together pressing the substrate against the inked plate thereby producing the printed substrate.

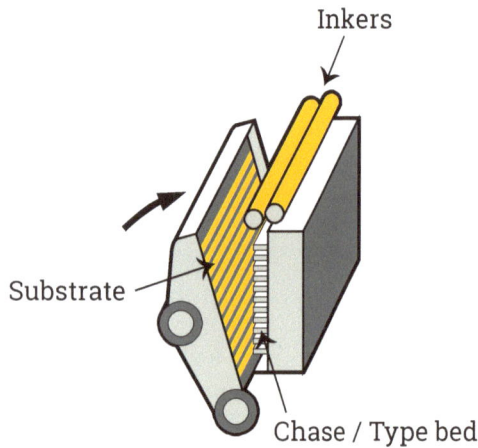

Figure 3.4 - Diagram of platen press *Source: 4impression*

LETTERPRESS SHEET-FED - FLATBED
The flatbed letterpress 'sheet-fed' press is a single color press. The flatbed configuration is constructed with a bed which holds the forme. This bed lies in a horizontal position whilst the rotating impression cylinder is in a fixed position and located over the flatbed. **(See Figure 3.5).** The bed moves back and forth
in a horizontal direction passing under the revolving impression cylinder which carries the substrate being printed. On the impression stoke, the traveling bed

Figure 3.5 - Principle of sheet-fed letterpress

and rotating impression cylinder are synchronised and the forme and cylinder travel together to transfer the image to the substrate. The impression cylinder then lifts to allow the bed to return to its first position; and the cycle is repeated. Flatbed cylinder presses are very slow running, with a maximum speed rate of 5,000 impressions per hour.

SEMI-ROTARY - LETTERPRESS
This type of semi-rotary press configuration uses a stop–start web feed system and can be fitted with multiple letterpress printing units, it can also be fitted with flatbed hot foil stamping units, flatbed embossing and flatbed die-cutting units and waste rewind.

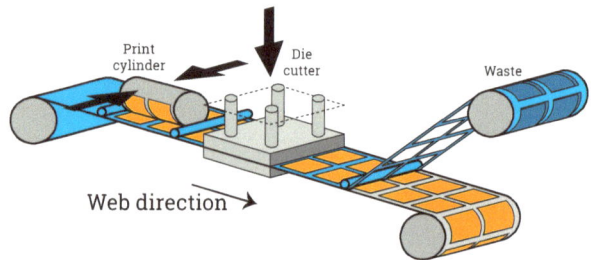

Figure 3.6 - Principles of semi-rotary stop-press letterpress

The semi-rotary print head differs from the full rotary press in two important areas. Firstly the printing head is positioned to print 'across' a stationary web **(See Figure 3.6)** and not 'down' the web as in a full rotary printing head.

The printing plate is located onto the plate cylinder and receives the ink film from a 'fixed' position inking unit. This cylinder rolls across the stationary web and transfers the image to the substrate. The cylinder then traverses back to the start position and fresh ink is applied to the plate as the web moves forward in readiness for the next printing cycle.

ROTARY LETTERPRESS - TRANSLATORY FEED

The letterpress technique is also used on intermittent ('translatory') fed presses. The press has full rotary printing units linked to a translator or reciprocating web feed **(See Figure 3.7)** and is a completely different configuration to the semi-rotary press. The

Printing Cycle 1

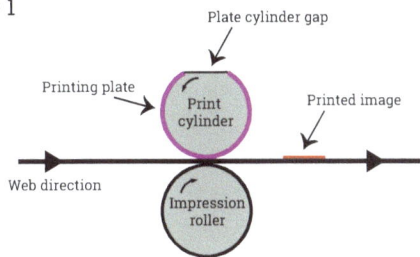

Printing Cycle 2

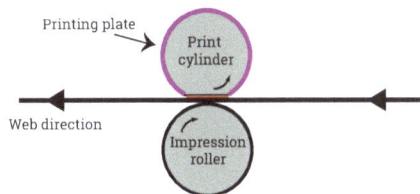

Figure 3.7 - Diagram of translatory feed press

press layout is the same as a full rotary web-fed press, but the difference lies with the web transportation system. The web travels through the press with a forward/backward reciprocating action and the printing cycle takes place as the web travels forward. The print unit has a 'fixed' cylinder repeat length meaning there are no variable size print cylinder changes required.

ROTARY LETTERPRESS - STACK PRESS
See chapter two for detailed information
The term 'Stack' describes the way in which the print

units are positioned on the press **(See Figure 3.8)**. The letterpress print units are 'stacked' on top of each other with sufficient space between each unit to allow for a drying unit to be located. The stack can be configured to provide twin stacks allowing 6 print units to be used.

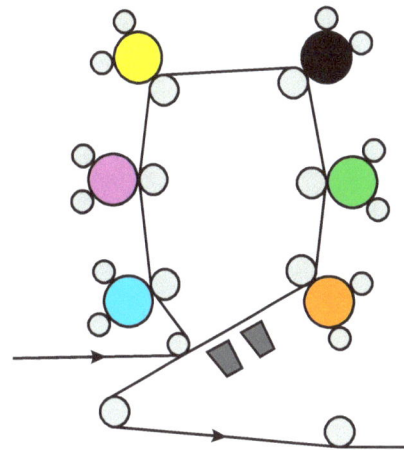

Figure 3.8 - Diagram of letterpress stack press. *Source: 4impression*

This type of configuration played an important part in the development and manufacturing of HAV (high added value) self-adhesive labels. Typical of this type of press are the Gallus R160 and R200. These 'stack presses' allow for the production of high quality print, on both filmic and metallised substrates, using combinations of the letterpress, flexographic and screen processes, linked to a UV curing/drying system. A short web path and easy makeready makes this type of press suitable for short or long run work.

ROTARY LETTERPRESS - CENTRAL IMPRESSION (CI)
This configuration of label press features a single large diameter central impression cylinder with letterpress printing heads located around the CI cylinder **(See Figure 3.9)** and the drying units positioned between each print head. These presses can be configured

with up to six print units positioned around the CI cylinder. The CI cylinder has a synthetic rubber surface which cushions the impression between the printing plate, the substrate and the impression cylinder, ensuring there is no metal to plate contact between the CI cylinder surface and the printing plate, which could lead to damage to both the printing plate and the CI cylinder.

Figure 3.9 - Central impression label press

TYPES OF LETTERPRESS PLATES

The letterpress plates used in the label industry comprise of three types; zinc/magnesium plates, polymer plates, and steel backed plates.

The zinc/magnesium plate is a ridged flat plate and is used in presses with a flatbed configuration. A chemical etching process is used to create the relief image.

The polymer plate is the plate most widely used in the label industry and the most suitable for full rotary presses. The plate is a film backed flexible plate, which is mounted on to the plate cylinder using accurately calibrated double sided adhesive film tape.

The steel backed plate is a polymer structure fused onto a thin steel shim. This type of plate can be supplied flat for mounting onto a magnetic base or in a semi-rotary format, which has punched locating holes positioned at the leading and back edges of the plate. These locating holes correspond with pins on the print cylinder configuration used on semi-rotary intermittent feed presses.

IMAGING THE LETTERPRESS PLATE

The imaging of photopolymer plates is done by exposing the polymer to a UV light source through a film negative. The film and plate are positioned in contact with one another and placed in a vacuum sealed unit to ensure full and even contact.

Exposure units may be flat or rotary. If the plate is for flatbed application then the exposure is done flat, but if the plate is for semi-rotary use then it is done in a rotary unit. This is necessary because allowance has to be made for the distortion factor that is created by the curvature of the printing cylinder.

UV light is then applied and the image area is polymerised and hardens leaving the unexposed area still soft, which can then be removed.

The washing/brushing process is done in a washing unit and the removal of the non-image area can be achieved with direct sprays, pads, or a rotating brush and sprayed with clean warm water. The brush must be soft to avoid damaging the plate surface. The plate is then dried using warm air and a second post cure exposure is given to the plate to ensure the polymer is fully hardened. **(See Figure 3.10).**

CTP (COMPUTER TO PLATE) IMAGING

Letterpress plate imaging is also done using CtP (Computer to Plate) direct ablation.

This is a process which uses specially formulated letterpress plates coated with a black ablation layer which means that the letterpress plate can be imaged directly via a digitally driven laser which is controlled from a computer file. The same laser systems are used for the imaging of a digital flexo plate.

The hard, metal-reinforced letterpress plates are coated with a black ablation (LAMS) layer that is evaporated by laser and removed by suction from the parts to be printed. The plates are then cured with UV light, washed out, dried and post-exposed. **(See Figure 3.10).**

If the press dot gain is established then the gain can be calibrated into the file and the plate image can be adjusted accordingly. This facility is of considerable benefit when linked to good impression control, allowing the printing of high quality process color reproduction.

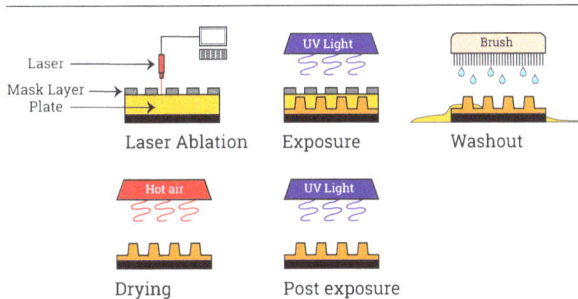

Figure 3.10 - CtP Letterpress plate making sequence

MAINTENANCE PROGRAM

A press that is not regularly maintained will not give the optimum printed image. A regular maintenance program therefore is very important. Bearings, print cylinders, gears etc. do wear and it is advisable that replaceable parts are held available so they can be changed when required without any delays being experienced. Presses are subject to stresses and strains caused by the use of heavy tooling units, continuous running and poor maintenance.

Cylinder bearings in particular have a limited life and a program of planned maintenance for the cylinder bearings should carried out, including cleaning, checks for wear and adequate lubrication.

LETTERPRESS PRINT CYLINDERS

In the flexographic, letterpress and litho processes the printing plate is located on the print cylinders. The cylinder needs to have accurate and even contact with the inking rollers and the surface of the substrate. The print cylinders should run perfectly true with an accuracy of ±0.025 mm ensuring that the pressure on the adjacent rollers is constant. Accurate measurement should be made when the press has been running for a short period to allow the running parts to warm up.

CYLINDER GEARS

The care and condition of the print cylinder gears is very important. When the gear is manufactured it is intended to be meshed at a certain depth with the gear it is driving. This is called the pitch diameter.

When meshing the print cylinder gear with the impression roll gear, this pitch diameter is critical to accurate register. Too deep or too shallow a mesh will cause loss of register between one color and the next. Regular lubrication is preferred to intermittent applications, as this will maintain a constant film of oil and even out temperature fluctuations.

The print cylinder circumference is based on the print repeat length in inches, millimeters or the number and size of gear teeth. Allowance has to be made for the thickness of printing plate plus the mounting tape and also for the effect of pressure and any expansion in the print cylinder.

The calculation would be as follows;

A print repeat length of 12 inches (or 96 1/8' gear teeth) equals a circumference of 300.8 mm. The print cylinder must have a circumference that is smaller by 3.14 (Pi), to allow for two thicknesses of plate and mounting tape, 3.14 x 2 (1.7 + 0.3) mm and a small allowance for the effects of the printing pressure and thermal expansion, say 0.01 to 0.03 mm.

PLATE MOUNTING

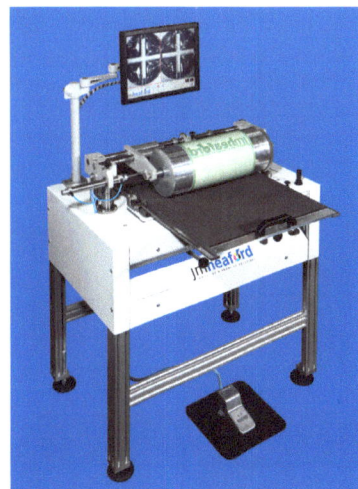

Figure 3.11 - Polymer plate mounting. *Source: JM Heaford*

Before mounting the printing plate the print cylinder surfaces should be thoroughly cleaned to ensure a clean, uncontaminated surface.

A contaminated cylinder surface will cause problems with the print quality.

Foreign particles trapped between the surfaces of the print cylinder, the mounting tape and the back of the printing plate and any grease or oily residue will affect the adhesion power of the mounting tape causing plate lift during the print run.

Double sided filmic adhesive tape is used to fix the printing plate to the printing cylinder. **(See Figure 3.11).**

These tapes can vary in thickness to allow for any slight variations in the diameter of the print cylinder and will compensate for any under-or oversized print cylinder. It is critical that the outside peripheral of each printing plate is exactly the same. If this is not achieved then accurate print to print register will be impossible.

It is strongly recommended that the same brand of tape should be used on each of a set of print cylinders.

Some tape manufacturers incorporate a thin layer of foam within the mounting tape which assists in smoothing out any small deviations in the plate cylinder.

MOUNTING FLEXIBLE PLATES

There are some basic principles that need to be followed when mounting plates. The mounting of flexible plates relies on manual skills, even though it is usually carried out using a plate mounting system equipped with mechanical and optical aids. The one rule that must be followed is to ensure that the plate is in the correct position at the first attempt of mounting. This involves a careful check to ensure that the registration lines on the leading edge of the plate are correctly in-line with either the optical system on the mounting system or alternatively the engraved grid on the print cylinder. These engraved lines run both horizontally and circumferentially round the print cylinder. Generally speaking the plate is usually mounted in the center of the cylinder and it is recommended the center of the plate is marked and the mark located with the center of the cylinder.

It is important that a little extra time is spent in ensuring that the leading edge of the plate is in the

correct position before completing the full mounting, as any removal of the flexible plate to 'try again', can distort and stretch the plate, making it difficult to remount and correctly position it over the whole of its area.

LETTERPRESS INKS AND DRYING SYSTEMS

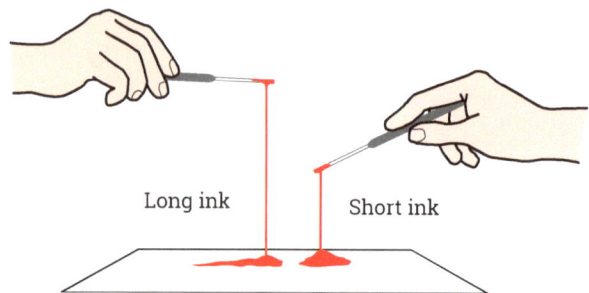

Figure 3.12 - Letterpress high viscosity short ink

Paste inks are used for both flatbed and rotary letterpress systems. The ink has a high viscosity and is therefore very similar to offset litho inks (often referred to as a short ink), **(See Figure 3.12)**. The letterpress plate has a hard surface and this ensures a good ink film transfer to the substrate being printed. The letterpress inking system uses an extended train of rollers which both rotate and oscillate from side to side in order to ensure that the thick paste ink film is evenly distributed prior to being applied to the printing plate. The majority of letterpress label presses are equipped with ultraviolet (UV) curing systems and the UV ink is especially formulated to cure rapidly.

IDENTIFYING THE LETTERPRESS PROCESS

All printed graphics are formed using a dot formation. The smaller and less dense the dots, the lighter the color. The larger and more dense the dot, the darker the color. This effect is called the tonal value. **(See Figure 3.13).**

Each of the printing processes has a particular characteristic which can be easily identified. A printed image produced by the letterpress process can be

Dots break up at 3.5%

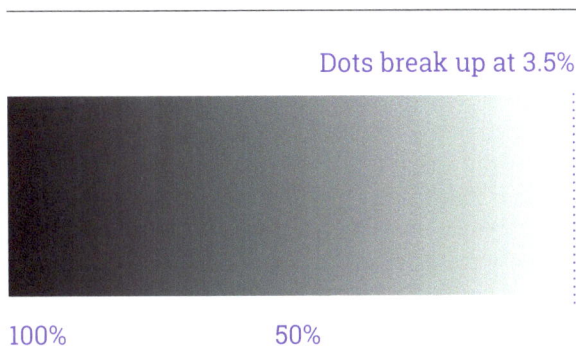

100%   50%

Figure 3.13 - Illustration showing dot break-up between 3-5%

identified by three characteristics:

1. The indentation on the reverse of the substrate that is created as a result of the pressure applied (impression) during the printing cycle.
2. The halo effect that can be created when there has been incorrect plate roller settings and too much printing pressure applied.
3. The 'hard edge' that is formed in the printed image as a result of dot break up, This is created by the inability of the letterpress process to print a dot below 3%.

ADVANTAGES AND DISADVANTAGES OF LETTERPRESS PRINTING

Advantages
- Good color density
- Extended plate life
- Good printing of type and solids and sharp edge definition
- No ink and water balance problems
- Suitable for sheet-fed and web-fed printing
- Good legibility of text even on uncoated papers;

Disadvantages
- Limited tonal reproduction
- Minimum printable dot (3-5%)
- Not suitable for thin filmic ie shrink sleeves
- Relatively high printing plate costs
- Poor tonal reproduction when compared with offset litho
- Relatively coarse screen rulings have to be used for tonal reproductions

LETTERPRESS PRINTING- SUMMARY

Although surveys into print process trends show flexography has overtaken letterpress as the leading narrow web process in the global market, letterpress is far from defunct. Indeed, letterpress is making a comeback as an option for low run, high complexity jobs, driven by new technology innovations. The cost efficiency of the letterpress process in particular is a key factor in stimulating its revival.

As has already been highlighted, letterpress has certain intrinsic advantages as a printing system compared to offset and flexo. To achieve exact color adjustment in flexo, repro has to be corrected, or colors need to be specially mixed. With letterpress, however, color can be adjusted during makeready.

Letterpress also has advantages versus the offset process in that running speeds can be increased from 10 m/min up to 60 m/min with no color change. This is difficult to achieve with offset because the ink/water balance is changing and this therefore creates waste.

Semi-rotary letterpress also has particular benefits with low tooling costs, allowing it to combine processes such as UV and solvent flatbed or semi-rotary screen, hot and cold foil stamping/embossing, lamination, reverse printing and flexo UV varnish.

Although digital presses are selling fast, there are still a lot of converters, even in mature markets, who still prefer the letterpress route. Semi-rotary technology in particular is ideal for small and medium jobs, because of its very fast changeover and the print quality is almost comparable to a digital press. Digital printing equipment, of course, is also very expensive.

TECHNOLOGY ADVANCEMENTS

Letterpress continues to advance technologically. There are still developments in the press sector – with semi-rotary, servo-driven systems, competing with digital on short runs of products, particularly those with complex converting requirements.

In pre-press digital letterpress platemaking is also undergoing major developments.

One of the most important developments has been ink trays with segmented blades allowing ink to be delivered to the plate in the exact quantities determined at pre-press. Exact inking profiles of jobs can be saved and recalled, allowing repeat jobs to be set up quickly and consistently

In developing markets, letterpress, particularly when matched with flatbed die-cutting, remains a powerful force, with converters often nervous about a shift to flexography because of the need to change to more expensive ink and plate systems.

Other technology developments have included new designs for printing sets and inking rollers, which help eliminate the effects of double imaging and ghosting, while new printing cylinder designs have delivered greater strike precision and increased uniformity of printed dots.

The development of magnetic sleeves also makes it possible to change between jobs very quickly and to carry out preparation away from the machine.

Other recent developments are occurring in letterpress platemaking with the introduction of water-washable plates. Digital letterpress plates are also emerging with digital stencil plates allowing conventional film processing to be replaced by digital mask ablation. Specially formulated letterpress plates coated with a black ablation layer can be imaged on the same system as a digital flexo plate.

The letterpress process is alive and well and backed by technology advancements. It still has a significant role to play in the label sector.

Chapter 4

The lithographic printing process

This Chapter explains the principles and workings of the offset lithographic printing process, as used in the label industry, and features both conventional offset and waterless offset printing.

HISTORICAL EVOLUTION

The word 'lithograph' historically means 'print from stone.' Developed as an economic means of reproducing artwork, lithography used printing stones produced from limestone, benefiting from their flat, porous surface on to which a greasy image was created and water applied to the stone to produce the print and non-print areas.

In 1796 the Bavarian author Alois Senefelder invented lithography, a printing process that used chemicals to create the image.

By the 1850s the process of chromo-lithography enabled the printing of up to twelve different colors, using a combination of dots and solid areas to create the printed image. This method of printing with stones dominated for the next sixty years, especially in the mass market. Thereafter, developments in the printing process concentrated on the speed of output and keeping manufacturing costs down, whilst maintaining quality.

In 1875 Robert Barclay patented the first rotary offset lithographic printing press. This press used transfer printing technologies that employed a metal cylinder instead of a flat stone.

The offset cylinder was covered with specially treated cardboard that transferred the printed image from the stone to the surface of the metal. Later, the cardboard covering of the offset cylinder was changed to rubber, which is still the most commonly used material today.

Lithography is commonly referred to as offset lithography or simply offset and is the most popular printing process for the printing of glue applied paper labels.

The use of the offset litho process on roll-fed self adhesive label presses increased in the early 1990s with the introduction of the Gallus T250, which is an intermittent feed, multi-process label press.

This was followed shortly after by the Nilpeter M300 full rotary, multi-process label press. One of the drivers for these developments was the need, within the label industry, for a print process that would produce high quality tonal graphics mainly for the personal care, wines and spirits and food markets. Driving this development was the need to produce self-adhesive labels that fully matched litho printed body, back and neck wet glue labels already used in the wines and spirit industry.

LITHOGRAPHIC PRESS CONFIGURATIONS USED IN LABEL MANUFACTURING

There are two formats of lithographic printing press used for the printing of labels: sheet-fed and reel-fed.

With sheet-fed litho the substrate to be printed is guillotine cut into sheets, creating a stack which is then loaded into the press and printed as a single sheet. This compares to a web-fed press where the substrate is fed from a master roll and then printed as a continuous web.

With litho printing each printing unit prints a single color and a press can be configured with the desired number of print heads as required. For example a 6 color press will print up to 6 individual colors or less if required.

Reel-fed presses are able to process a reel-fed substrate, which can then be cut into single sheets after the printing and converting processes.

The ability to print substrates from the reel is of great benefit, allowing the web press to process lightweight substrates (both filmic and paper).

This advantage derives from the principle of the process which means that the substrate face is only exposed to a flat (planographic) surface during the printing operation and is not distorted by contact with a printing plate which has a 'relief' image surface, as is the case with the letterpress and flexographic printing processes.

THE SHEET-FED LITHO PRESS

Within the label industry the sheet-fed press is used predominantly for the printing of wet glue applied labels. Individual sheets are held in a stack or pile and fed one at a time into the press. Delivery of the printed sheets is through the press 'delivery' system, where the printed sheets are placed into a stack or pile at the end of the press. The sheets can be then reprocessed or cut into single labels for product application.

The commercial print market is the largest user of the sheet-fed offset litho process with presses usually comprising of 6 print units plus a coating unit for overall or spot varnish requirements. Larger volume book and magazine production is generally printed on an 8-color press which is configured to print 4 colors on the front side of the sheet and 4 colors on the reverse of the sheet; this process is known as 'perfecting'.

Sheet-fed litho usage within the label industry falls into three categories:

1. **Wet glue labels.**
2. **In-mold labels**
3. **Self-adhesive labels printed in sheet form**

REEL-FED LITHO PRESS

The printing, embellishing and converting of self-adhesive labels is carried out on a web-fed press, also known as a reel-fed press. **Figure 4.1** shows a web-fed 4 color press commonly used in the label industry. This type of press prints onto a continuous web and not in a single sheet format, which means that the reel is placed on the in-feed section of the press, known as the unwind unit, and the printed and converted reel exits after unit 4 onto the rewind unit.

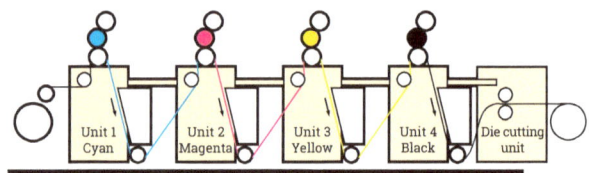

Figure 4.1 - Configuration of 4 color web-fed litho press

Figure 4.1 shows a 4-unit web-fed press printing in the process colors of CMYK, Cyan unit 1 – Magenta unit 2 - Yellow unit 3 and Black unit 4. The printing head is mounted on the top of each unit and the die-cutting station is typically positioned between unit four and the rewind unit.

One of the major differences between the sheet-fed press and the web-fed press is the control of web tension during the printing operation. Sheet-fed equipment does not require web tension control as each sheet is individually controlled by a system of gripper cylinders which ensure very accurate print to print registration between each color.

With web-fed presses the control of the web tension is critical and the web control system must ensure that there is no movement of the substrate as a result of substrate shrinkage or stretch, which in turn can create mis-register of the colors being printed. Modern web-fed presses use very sophisticated digital web control and print register systems, using web sensors linked to step motor technology and the controlled use of ink drying and curing systems that minimise the amount of heat exposure to the substrate during the printing process. This limits the shrinkage and stretch factors, particularly when filmic substrates are being printed.

OFFSET - CYLINDER REPEAT LENGTH

Litho printing in the label industry has experienced some difficulty in adapting to the varying size requirements of printing labels in a reel-to-reel format.

Offset lithography is ideal for sheet printing and reel printing with a fixed repeat length. However, because of the cylinder configuration of an offset unit, the plate cylinder sizes are restricted. A number of successful attempts have been made to modify the system to make it more adaptable to the wide variations required when printing labels in all sizes.

Consecutive jobs are rarely of the same repeat length and therefore require a time consuming changeover of the cylinders holding the printing plates.

Some press manufacturers have introduced systems which appear to effectively eliminate the need for extensive changeover of rollers and gears in order to accommodate the wide range of printing lengths required.

The semi-rotary systems are one way of eliminating the changeover of print cylinders and gears. However, these systems are much slower running than the modern full rotary systems.

The semi-rotary press design uses a forward and backward movement of the web rather than a continuous forward movement, thus taking material through and then bringing it back to compensate for the various repeat lengths required by each job. The web is automatically brought back after each printing cycle, controlling the gap between images and ensuring that the printed image remains in register between one print station and the next.

THE PRINCIPLES OF THE OFFSET LITHO PROCESS

Modern offset litho printing uses a printing plate made from thin grained metal sheet that carries the photographically developed image.

The process works on the principle that oil and water do not mix. If an imaged area on a plate **Figure 4.2** is damped with water then the image area will reject the damping, but the non-image area will accept the damping **(See Figure 4.3)**. If ink is

passed over the plate, the image area will accept ink, but the damped non-image area will reject ink. **(See Figure 4.4)**. The inked image on the plate is now ready for transferring onto the offset blanket and then to the substrate.

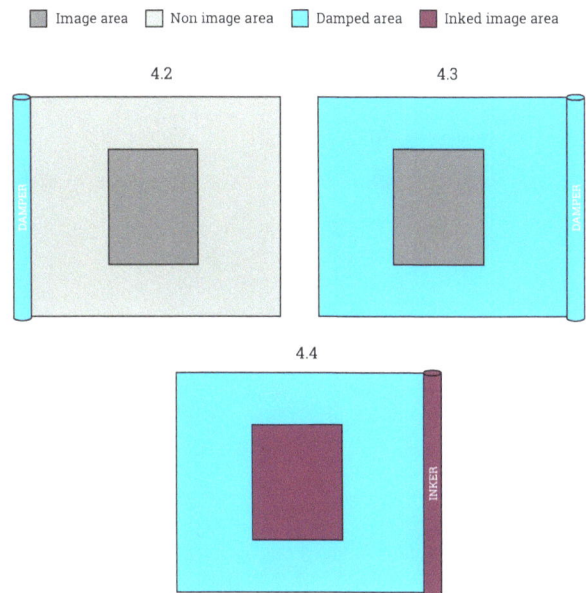

| ■ Image area | □ Non image area | ■ Damped area | ■ Inked image area |

Figure 4.2, 4.3, 4.4 - Principles of the offset litho process.
Source: 4impression

You can see the non-printing area in light grey, the print area in dark grey and the damper roller in blue.

The inking roller, in red, passes over the plate. The oil based ink is rejected in the non-print image area by the film of damp, but is accepted by the damp free print image area.

The print area is inked and the image transferred onto the offset blanket and then 'offset' printed onto the substrate.

THE PRINTING CYCLE

Figure 4.5 shows the cycle of damping and inking on a litho printing unit, with the plate cylinder, damper rollers and inking rollers in position.

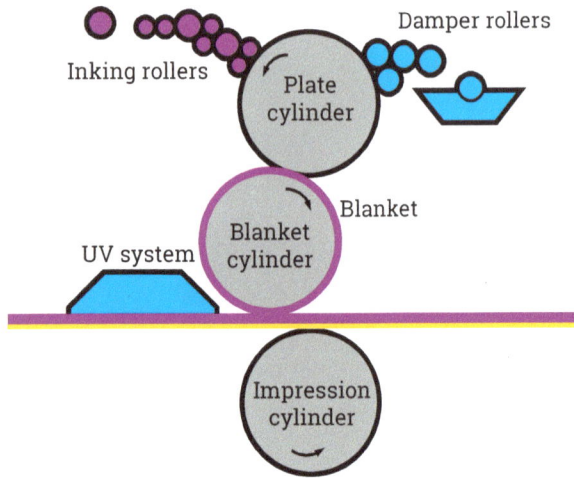

Figure 4.5 - Diagram of conventional offset litho unit.
Source: 4impression

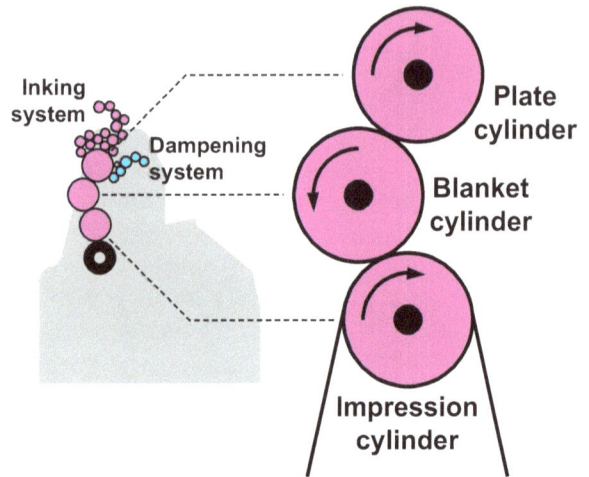

Figure 4.6 - Location of plate, blanket and impression cylinder (4impression)

This graphic illustrates the rotation of the plate, blanket and impression cylinders and shows the roller trains used for the damping and inking of the printing plate.

The inked image (purple) is then transferred onto the rubber offset blanket and then under pressure onto the substrate (yellow). The ink is then dried or cured.

The volume of the ink film to the plate is controlled via the ink reservoir (called the ink duct) and is regulated through a system of rollers (called the roller train). The roller train ensures that the ink film is correctly regulated and consistent on every revolution of the plate cylinder.

THE OFFSET UNIT
Figure 4.6 shows the location of the three cylinders and the inking and damping roller systems within the printing unit.

On the right side of the illustration the cylinder configuration of impression cylinder, blanket cylinder and plate cylinder can be seen in more detail.

In printing mode the inked image is transferred from the plate cylinder onto the offset blanket (positioned on the blanket cylinder) and the pressure between the blanket and the impression cylinder transfers the printed image to the substrate.

The control of the damping process is critical in achieving the correct print quality. Any imbalance associated with insufficient damping will allow the ink film to contaminate the non-image area and this will result in ink being deposited in the non-image area thereby creating a scumming or catch-up.

Scumming can be adversely affected by the ambient heat around the press and also the heat generated when running the press at speed.

Modern damping units use a very controlled system of alcohol damping, including the use of chemical additives to produce the optimum damping control. This solution is called the 'fount solution'.
Figure 4.7 Shows an offset printing unit located on a label press. The arrows identify the three cylinders on a reel-fed self-adhesive label press, showing the plate cylinder, the blanket cylinder and the impression

Plate cylinder
Blanket cylinder
Impression cylinder
(beneath Blanket Cylinder)

Figure 4.7 - Photographic image of actual offset unit.
Source: Rotatek

cylinder positioned below the blanket cylinder.

The litho process produces an evenly inked image and although the process applies only a thin film of ink, the image is sharp with no edge blemishes and thereby delivers an excellent printed result.

The reproduction of fine detail, including tonal images is excellent with only the digital process and the gravure process equal to it in print definition.

THE OFFSET BLANKET

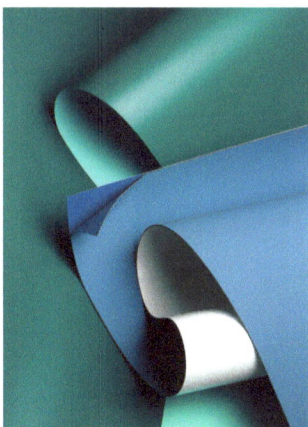

Figure 4.8 - Blanket material used in the litho process.
Source: Trelleborg

To reduce wear (abrasion) of the plate, the image is transferred to the printing substrate via the offset blanket which is positioned on the blanket cylinder.

The blanket has a self-adhesive backing and the leading edge of the blanket is positioned and wrapped around the cylinder, the surplus tailpiece is then cut off to ensure that there is a smooth area between the two edges of the blanket.

Offset blankets are made of synthetic rubber and are available with a variable shore hardness allowing the press operator to choose the correct blanket hardness. It is important that the release factor of the blanket is correct to suit the substrate being printed, so that the inked image is fully transferred to the substrate on every revolution of the printing cycle.

Washing/cleaning of the offset blanket is very important as this ensures that the image is cleanly transferred on every print cycle. Ink residue and fibres from the substrate surface can contaminate the blanket and regular cleaning throughout the print run is recommended. Blanket cleaning is a manual operation but some litho presses are fitted with automatic blanket wash systems.

THE LITHO PRINTING PLATE

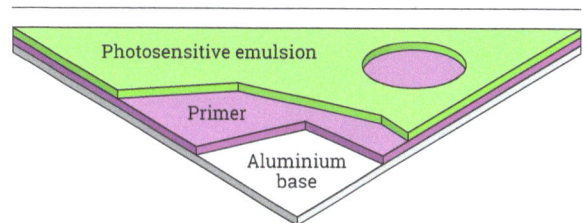

Photosensitive emulsion
Primer
Aluminium base

Figure 4.9 - Structure of litho plate. *Source: 4impression*

The image and non-printing areas of the printing plate are in the same plane i.e. a flat surface hence the term 'planographic' printing process.

The illustration in **Figure 4.9** shows the structure of a lithographic printing plate, which comprises of a aluminium base, a primer layer and finally a photo sensitive coating.

PLATE IMAGING
There are two types of method for imaging the printing plate;

1. Contact imaging where an imaged film is in direct contact with the print plate and given a timed exposure to establish the image.
2. CtP (Computer to plate) where a digital file is imaged directly onto the plate without the need for film.

CONTACT IMAGING

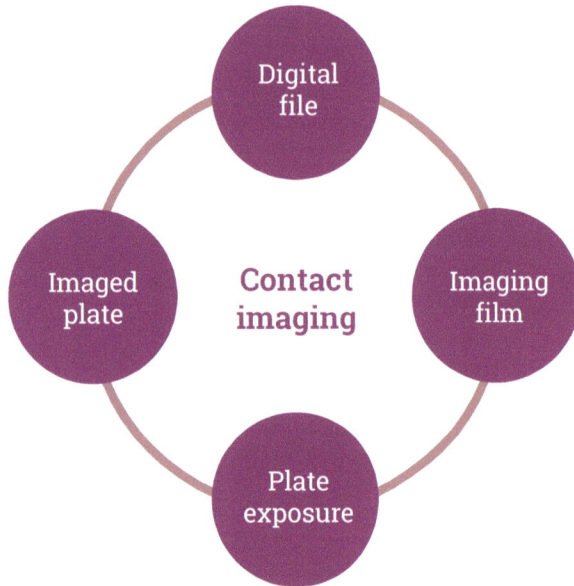

Figure 4.10 - Contact imaging sequence

Figure 4.11 - Printing plate with film negative

negative separated after exposure. The image is clearly visible and the plate is now ready for developing.

COMPUTER TO PLATE - IMAGING (CTP)

Offset plate imaging can also be done without the need for film originals using a process called CtP or 'Computer to plate'. (**See Figure 4.12).**

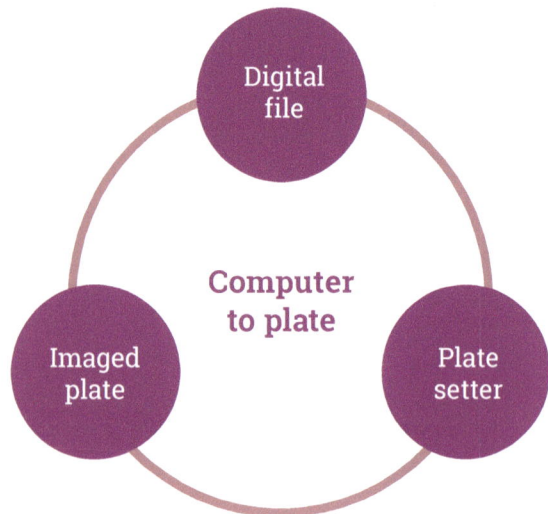

Figure 4.12 - CtP sequence

Although most imaging techniques use CtP (computer to plate) film based imaging is still in use. (**Figure 4.10).** The film negative or positive which is created from the digital file is placed in direct contact with the plate and exposed to a UV light source.

The image from the film is transferred to the printing plates using a photographic process. A measured amount of light is allowed to pass through the film negative thereby exposing the printing plate. On exposure a chemical reaction occurs that activates the ink receptive imaged area. The plate is then developed and the image is chemically fixed. The plate is then ready for positioning into the press.

Figure 4.11 shows the printing plate with the film

The digital file to be printed is transferred to a CtP device called a plate setter and the image is created using direct laser imaging.

After laser imaging, the emulsion that remains in the imaged area is removed leaving it ink receptive. CtP plates do not require chemical processing.

CtP plate imaging eliminates the need for film-plate exposure and chemical processing.

Once the litho plate has been imaged and checked, it is now ready for locating into the press prior to the printing operation taking place.

ON THE PRESS: PLATE CHANGE

The plate is first located into the plate cylinder with the leading edges of the plate positioned first. The plate is then wrapped around the plate cylinder and the trailing edge located, positioned and clamped. The operator then applies the correct tension to the plate to ensure that the position is correct and the image is in register.

PRINTING INKS

There are two types of litho printing ink systems:

Conventional oil based and UV (ultra violet) oil based. The UV system offers some considerable advantages over the conventional system and these advantages will be explored later in the chapter.

In **Figure 4.13** you can see the consistency of the litho inks and press operators can easily handle the inks with a standard palette knife making color mixing /matching much easier.

Figure 4.13 - Consistency of litho ink (short ink). *Source: Flint Group*

INK DISTRIBUTION CONTROL VIA THE INK DUCT

The duct roller delivers ink to a drop roller which moves between the duct roller and the distribution rollers. On some litho presses certain distribution rollers are chilled/cooled to assist in controlling the temperature of the inking rollers which can generate a lot of heat within the roller train, as a result of the high speeds at which the rollers turns. The combination of heat and speed can also create ink 'flying' which produces a spay mist of ink.

The purpose of the distribution rollers is to evenly distribute the ink film which is transferred to the plate via the plate inking rollers also known as 'forme' rollers.

Over a number of years the efficiency of the ink roller train has improved and press manufacturers have done considerable development work in providing an inking system that ensures that on every print cycle the plate receives the exact ink film required. These developments include varying the roller diameters and adjustable roller oscillation.

One of the most common faults experienced by the pressman is called 'ghosting'. This problem is most evident when printing solids which incorporate smaller reversed-out image areas. The problem manifests itself by printing a ghost image onto the solid areas which is very difficult to eliminate. This issue is usually attributed to inadequate rolling power and can be further aggravated by incorrect ink viscosity.

CONTROL OF THE INK FLOW

Control and adjustment of the volume of ink which is applied to the printing plate is done via the ink duct adjustment.

The flow of ink to the inking rollers can be increased or decreased by adjusting the segmented plates via the duct keys which form part of the ink duct and are positioned against the duct roller. This allows the ink film density to be varied across the width of the web being printed. The image being printed may for example comprise of solid content in some areas which require more ink, and other areas with text only that require less ink.

Adjustment of the individual keys which control the

flow can be done manually or through a remote keyboard.

ON THE PRESS: REMOTE INKING CONTROL

Modern offset litho presses are fitted with remote control inking systems.

These systems allow the press operator to make very fine adjustments to the ink flow during the print run. These color corrections are controlled from a master station and eliminate the need for the operator to move from print unit to print unit to make manual adjustments to each individual color.

Some litho presses, both sheet-fed and web-fed, use a closed loop color monitoring system for the control of the ink film being printed. The consistency of a specific color throughout the print run relies on maintaining a very accurate ink film weight.

These inking control systems offer major benefits by giving remote inking control through digital data, automated color adjustment during the press run and also quality control via on-press scanners that check both label content and color.

On-press scanners are used to take spectrophotometry and densitometry readings of the printed web or sheet. These readings relate to color density and dot gain* and the data is then cross referenced to the correct color specification. Any variation in color or dot gain is automatically adjusted on the press to ensure that the color specification and print sharpness is correctly matched with the print specification.

*DOT GAIN

Dot gain is the term used when the dot size prints larger than the correct specification, it is a print characteristic in which the size of the half-tone dot changes as a result of platemaking and also the printing process used.

When dots are transferred from a film to a plate, they will tend to grow in size during light exposure and the ink film which is transferred from the plate to the label substrate can increase in size. Fluid inks and compressible plates tend to increase the dot gain but this can vary according to press and substrate being used.

All printing presses will produce a certain amount of gain as a result of the engineering tolerance which occurs in the press manufacturing. Dot gain can also stem from ink viscosity variances and from incorrect pressure between the plate and impression cylinder (flexo and letterpress) or between the plate and blanket cylinder or the offset blanket and the substrate (litho) and it is important that these pressures are set correctly in order to reduce dot gain to a controlled minimum.

Dot gain is largely predictable and allowance for dot gain can be accommodated at the repro stage provided specific plate, substrate and press dot gain information is known.

Typically, the printer will undertake a press fingerprinting analysis to obtain dot gain information, It is very important that the press dot gain characteristics are identified by carrying out a dot gain test to establish the gain for each individual press, this reading can then be applied to the digital data used in the preparation of the image file and the dot size is reduced on the film to compensate for the gain that will take place during platemaking and the printing process.

DRYING SYSTEMS

There are three types of drying systems used within the commercial and self-adhesive printing industry – oxidisation, Infra Red and Ultra Violet, known as UV.

The modern self-adhesive press is fitted with inter-deck drying units. This allows for each individual color to be cured/dried before printing the next color. This is called printing 'wet on dry' and eliminates the problems which can be encountered when printing conventional litho inks 'wet on wet'.

DRYING SYSTEMS: OXIDISATION - NATURAL EVAPORATION OF OIL/SOLVENT

This is the method of drying the oil and solvent component used in these oil based ink systems by natural evaporation and where the drying process of these inks is not accelerated by using infra-red or hot air.

Oxidisation is a slow process because it relies on the absorbency of the substrate and exposure to the ambient air conditions

This method is not suitable for the web-fed

presses used in the self-adhesive industry.

DRYING SYSTEMS: INFRA-RED/HOT AIR

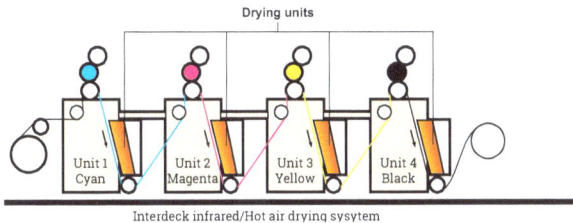

Figure 4.14 - Infra red drying on litho press. *Source: 4impression*

Infra-red is the method of accelerated drying which uses either hot air or direct heat or a combination of both hot air and heat. **(See Figure 4.14).**

Drying by infra-red is only suitable for oil, aqueous (water-based) and solvented inks. This type of drying system is compatible with many of the web-fed presses used in the self-adhesive industry.

Infra-red drying is not suitable for filmic materials as the heat that is generated creates problems with shrinkage and stretching of the filmic substrate, in particular lightweight unsupported film.

INTERDECK UV

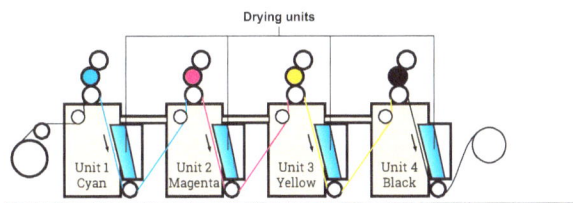

Figure 4.15 - Interdeck UV drying on litho press. *Source: 4impression*

The drying or curing of UV printing inks takes place through the reaction of the ink chemistry to a strong ultra violet light source.

UV inks are 100% solid system and do not contain solvents that must be evaporated during the curing phase.

Figure 4.15 shows the position of the UV units within the printing press. The curing systems used can be run at very high production rates, as the curing of the UV ink takes place rapidly. Drying speed however, can be affected by the color and density of the ink film and the intensity of the UV light source.

The increasing use of UV inks in the self-adhesive industry has made a considerable difference to the types of substrates which can be printed and converted. As the ink cure operates at low temperatures (known as 'cold cure'), one of the major benefits to the self-adhesive manufacturer is the facility to print and convert filmic and metallic substrates.

UV CASSETTE SYSTEM

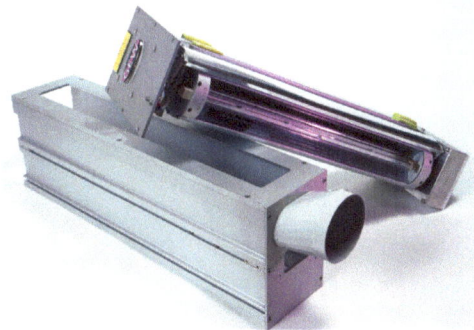

Figure 4.16 - Typical UV Curing System. *Source: GEW*

The modern UV system is far easier to operate and maintain than the early systems **(See Figure 4.16)**. The power of the ultra violet intensity can be adjusted to suit a particular ink curing requirement and the controlled removal of the unwanted heat generated by the infra-red ensures that the web is only exposed to a minimum amount of heat. This 'cold cure' UV makes the printing of light filmic substrates much easier.

TONAL VALUES

One of the key advantages offered by the litho

process is unrestricted tonal values... but what is meant by the term tonal values?

All graphic content is printed using a dot formation. The lighter the dot area, the lighter the color. The more dense the dot formation the darker the color. **(See Figure 4.17).**

This variation of dot density and size gives the various tonal values required to form the printed tones.

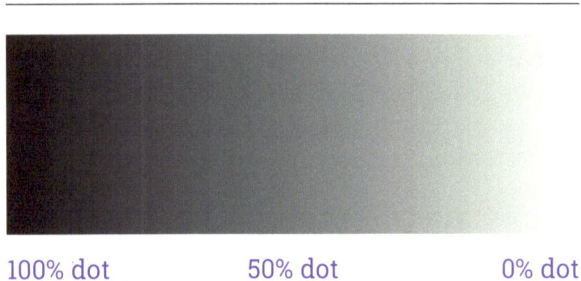

100% dot   50% dot   0% dot

Figure 4.17 - Transition from solid to infinite dot (vignette). *Source: 4impression*

IDENTIFYING THE LITHO PROCESS

It is necessary to use a magnifying glass to identify what type of printing process has been used on a particular graphic reproduction.

Each process has a particular characteristic which can be easily identified. As explained earlier the printed image is made up of differing densities of dot. The closer the dot formation the denser the color and the lighter the dot formation the lighter the color.

The offset litho process does not have the problem of limited fine dot reproduction. This means that the reproduction of very fine tonal values such as subtle tones and vignettes can be produced without difficulty, as the litho process does not have the same dot break-up experienced by other printing processes.

ADVANTAGES AND DISADVANTAGES OF THE OFFSET LITHO PROCESS

Summary of the advantages and the disadvantages offered by the litho process:

Advantages

- There are no tonal restrictions. The litho process has very clean edges on very fine line-work and printed tones that do not break up, allowing an extremely fine dot to be printed.
- Offset printing delivers excellent graphic and photographic reproduction.
- CtP imaging of the printing plate gives a fast and consistent plate output.
- Offset printing is very cost effective particularly when producing high quality work.
- Very high production speeds and fast makeready can be achieved.
- The litho process is ideal for multi-process combination presses. This is a major advantage to the self-adhesive industry as it allows the printer/graphic designer to choose the print process best suited to a particular graphic result.

Disadvantages

- Damping and inking imbalance on the press can be experienced, resulting in print problems in both the non-image and image areas
- Offset litho printing equipment can be expensive compared to other print processes used in the self-adhesive industry
- Dependant on the types of inks being used, litho plates have a shorter plate life compared to other print processes and the plate life can be affected by the use of metallic inks which are very abrasive.
- The litho process cannot print the heavy film weights which can be achieved by some of the alternative printing processes. An example of this would be the very heavy ink film weights that can be printed by the silk screen process, compared to the litho process.

WATERLESS OFFSET
THE WATERLESS PROCESS

Waterless offset, as the term implies, is an offset litho process that does not use a damping system. The pre-sensitised printing plate used does not require a dampening process and therefore can overcome some of the ink and water imbalance problems that can occur with conventional offset litho printing. **(See Figure 4.18).**

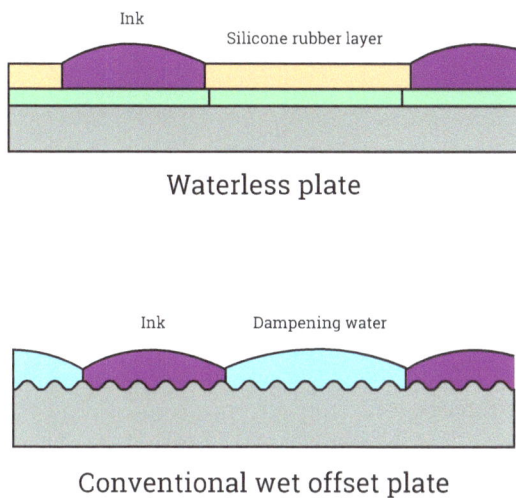

Figure 4.18 - *Comparison of waterless and conventional offset plate structure. Source: 4impression*

The rejection of the oil-based ink in the 'non-print' area is provided by a pre-sensitised printing plate that is coated with an ink repellent silicone rubber layer. The 'print image' area is slightly recessed and formed from an ink receptive polymer surface.

THE WATERLESS LITHO - PRESS CONFIGURATION

Let's take another look at the layout of a 'conventional' offset printing unit. **(See Figure 4.19).** As can be seen this unit has a standard damping system and an inking and distribution roller train. The temperatures of the surface area of the printing plate are not critical to achieve a perfect printing result and therefore the unit does not require a sophisticated cooling facility for the inking system or the printing plate.

The waterless printing unit, **(See Figure 4.19)** however, does not have a damping system and is fitted with a cooling system which supplies cooled

fluid via the inner section of the individual rollers used in the inking system. Chilled air is blown onto the surface of the printing plate to ensure that the ink

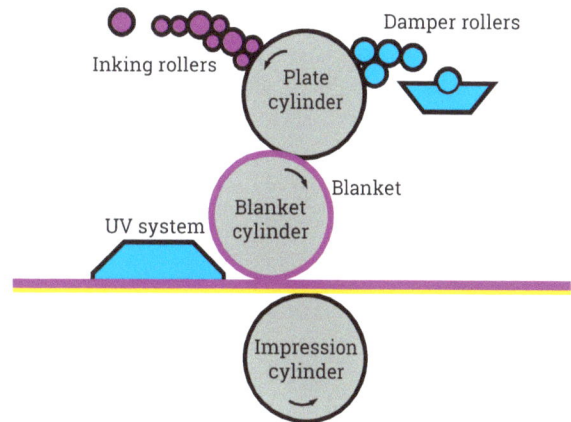

Figure 4.19 - *Conventional litho configuration. Source: 4impression*

rollers and the surface temperature of the printing plate are controlled. If the surface temperature of the printing plate exceeds 25 degrees centigrade the

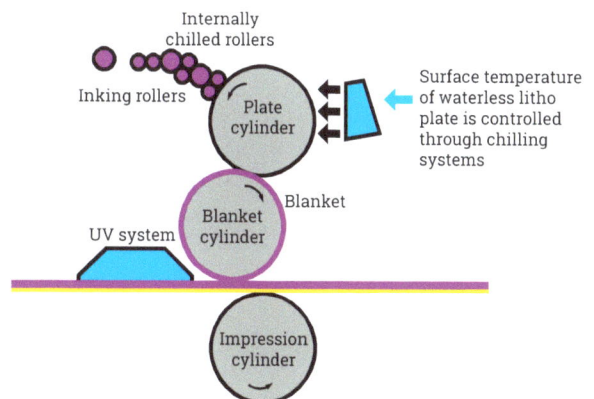

Figure 4.20 - *Waterless offset configuration. Source: 4impression*

siliconised rubber layer will breakdown. This will produce a similar imperfect image to that associated with insufficient damping, which results in the ink contaminating the non-image area (referred to as scumming or catch-up).

It is critical that the cooling system for the inking rollers, which can be chilled water or chilled oil, is capable of maintaining a consistent temperature below 25 centigrade.

WATERLESS PLATE IMAGING

The waterless lithographic plate is made up of an aluminium base which is covered with a primer and a photosensitive polymer layer and finished with a siliconised rubber layer and finally a protective layer of transparent film.

The plates can be imaged with positive or negative film. The film is brought into direct contact with the printing plate and exposed to ultraviolet light, in the same way as 'contact Imaging' in conventional litho plate imaging.

The UV exposure fuses the silicone to the photopolymer in the non-image areas creating a siliconised non-print area that will reject the ink. The next stage is the developing process when the silicone layer is washed away from the plate in the 'image areas', thereby allowing the ink to adhere to the image.

Plates can achieve screen resolutions of up to 800 lines per square inch.

WATERLESS AND CONVENTIONAL LITHO - THE PROCESS DIFFERENCES

Below is a summary of the key differences between waterless versus conventional litho:

- No damping is required with the waterless process
- The waterless plates use water-repellent silicone rubber coating, eliminating the need for a damping system.
- The silicone rubber layer in the image area is removed in the developing process.
- Special inks are required when printing with the waterless process and these inks are suitable for UV curing systems.

- The press requires a suitable cooling system to maintain a low temperature for the inking / distribution rollers and printing plate surface

ADVANTAGES AND DISADVANTAGES OF WATERLESS LITHO

Advantages
- It gives better tonal values and shadow contrasts
- There is improved color density and color consistency as a result of the absence of damping and therefore no water and ink imbalance issues
- Sharper dot reproduction
- Waterless litho gives an excellent printing result when using metallic inks

Disadvantages
- The critical requirement for accurate temperature control of the plate surface and the roller inking system,
- The surface temperature of the printing plate must not exceed 25 centigrade.
- Requires special inks

OFFSET LITHO PRINTING - SUMMARY

One of the key advantages offered by the litho process is the unrestricted tonal values it can achieve.

It is this ability to deliver a consistently high image quality that has seen the process dominate the wet glue sector for many years. Large volumes of litho labels produced on sheet-fed presses are used to decorate containers in high-value sectors such as wines, spirits and premium beers.

Litho has developed as a significant process in the self-adhesive label sector where the benefits of the process have been used for designs that require the reproduction of subtle tones and vignettes.

The litho process is ideal for multi-process combination presses where its advantages are often used in conjunction with other printing processes. Offset change cassettes systems allow the process to be inserted into the press with other processes. An option for converters with platform flexo presses is to use interchangeable offset cassettes in the flexo press line.

Web offset has proven particularly popular with converters making the transition from sheet-fed to roll-fed labels. Reel-fed printing offers many cost benefits over sheet-fed production, where a sheet may need to be reprocessed over and over again to obtain a result that can be achieved in a single pass with reel-fed.

TECHNOLOGY DEVELOPMENTS

The latest developments in web offset technology are making the process more attractive for shorter runs of labels and increasing in-line options available to converters.

A key development that is encouraging this trend is the use of sleeves for plate and blanket cylinders.

The cost of the aluminum-based sleeve system is believed to be around a fifth of the price of an offset cassette. The sleeves are mounted on hydraulic expansion shafts, allowing fast changes between jobs. This standardised system allows print and converting technologies to be switched quickly on the offset platform.

Known as variable-size offset printing (VSOP) it is very easy to change the printing length without changing the complete offset insert. The biggest advantage of this fast and uncomplicated changeover is that it is very cost effective, especially when many different printing repeat lengths are required.

VSOP is a cost-effective alternative to conventional flexo and rotogravure printing on small-to mid-size label runs where the quality and the pre-press costs are tipping the scale in favor of offset printing.

Chapter 5

The flexographic printing process

The origins of the flexographic printing process derived from the use of rubber stamps which were manufactured using plaster moulds impressed with lead type to create the image area. It was the Mosstype Corporation who first developed the use of rubber plate-making to be used in both the flexo and letterpress printing processes.

In 1890 the first flexo press was built in Liverpool by the English company Bibby Baron and Sons. Other presses which also used rubber plates and aniline* inks were developed in Europe, with Germany manufacturing the majority of flexo/aniline presses.

Originally, the inks used for flexo were aniline dye inks and the name aniline printing persisted. In the United States the aniline process was used extensively in the food packaging sector, but in the 1940s the use of aniline inks was banned in food packaging. Safer coloring agents were developed, but the aniline processes still carried a bad reputation and as a result the print sales which used this process declined. Franklin Moses in 1951 started a campaign to change the name of the process from aniline to the flexographic process which was subsequently renamed 'flexography.'

*Aniline - definition
*Aniline is a colorless, oily, partially water-soluble organic compound derived from nitrobenzene.

FLEXOGRAPHIC PRINTING IN LABEL MANUFACTURING
Flexographic printing was for many years considered an inferior printing process and serviced the lower end of the label market. Labels requiring higher quality were generally printed using the offset litho or letterpress processes.

In the last three decades considerable progress has been made in the manufacture of highly engineered flexo presses and developments in printing plates, anilox rollers and printing inks now makes flexo the leading process for the manufacture of self-adhesive labels.

The introduction of UV ink curing has had a big impact on the flexo process. The switch from aqueous and solvent based inks to UV cured ink in particular has improved print quality and made it easier to print filmic and metallic substrates and has reduced the usage of solvents.

TYPES OF FLEXO LABEL PRESSES
There are three configurations of flexo presses used in the label industry: the stack press; the common impression press; the in-line press.

All these presses can be equipped with solvented, water-based and UV drying systems dependent on the press specification. The press configuration most widely used for label manufacture is the in-line press.

THE PRINCIPLE OF THE FLEXOGRAPHIC PROCESS

Flexography is a relief or raised image printing process using the same principle as letterpress. A flexible rubber or photopolymer plate is mounted on to the plate cylinder using a filmic double-sided adhesive tape. The plate image is inked with a liquid ink which is transferred from the anilox roller, direct to the surface of the image area and then printed onto the substrate, using a very light controlled pressure (impression).

The basic flexo printing unit comprises of an ink tray or duct, the ink applicator roller, the anilox roller, the plate cylinder and the impression roller. **Figure 5.1** illustrates the layout of a standard flexo unit. The applicator roller runs in the liquid ink held in the ink tray. The ink can be manually poured into the ink tray or alternatively pumped into the tray using a circulatory system to ensure that the ink viscosity is maintained. The applicator roller applies ink to the anilox roller and the pressure between these two rollers can be adjusted to increase or decrease the ink film to the printing plate.

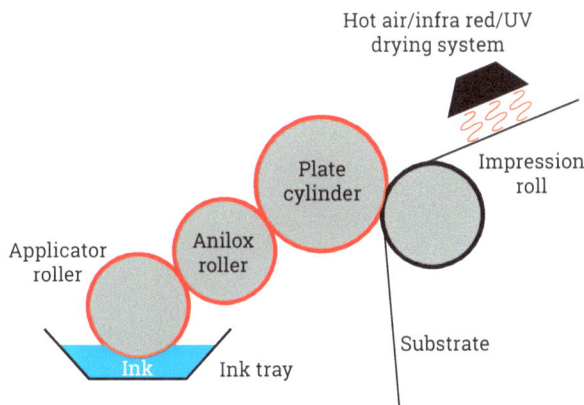

Figure 5.1 - Basic flexo unit. *Source: 4impression*

THE REVERSE ANGLE DOCTOR BLADE

The need for ink film accuracy and consistency has led to the development of a number of modifications to the flexo system which have improved the control and accuracy of the ink film delivered to the plate. The removal of excess ink from the anilox roller was greatly improved by the use of a reverse angle doctor blade unit **(See Figure 5.2)** and similar in principle to the type used in the gravure process. This development gives greater control over the ink film offered to the printing plate regardless of the press running speed.

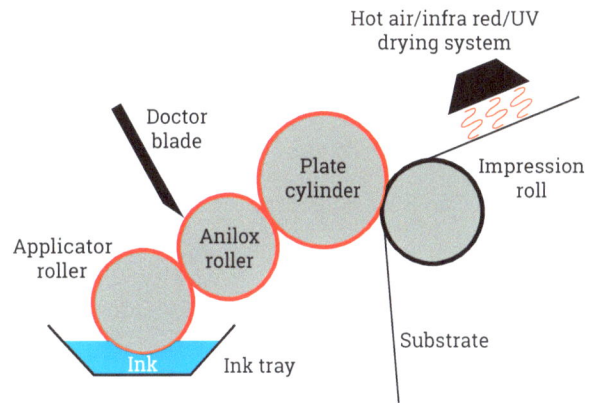

Figure 5.2 - Location of reverse angle doctor blade. *Source: 4impression*

THE CHAMBERED DOCTOR BLADE SYSTEM

The modern flexo press now uses a very accurate and efficient method of anilox doctoring; this is called a chambered doctor blade system. **(See Figure 5.3 and 5.4).**

The applicator roller is not required for a chambered system and the doctor blade unit has two blades which are held in the ink chamber unit. The blades are positioned above and below the center point of the anilox roller and maintain a constant contact with the anilox face, thereby creating the enclosed 'chamber' which holds the ink. The unit operates under a low pressure to ensure an even contact with the anilox face. The ink levels within the chamber are maintained using a circulatory pumping system.

THE ANILOX ROLLER

Anilox rollers **(See Figure 5.5)** are one of the most important elements of the flexo process. The full surface of the anilox roller is engraved with recessed cells and each cell holds a specified volume of ink thus ensuring that a consistent ink film is delivered to the printing plate.

Each individual anilox roller is made up of a steel shaft with end bearings which hold a copper sleeve. The cells are engraved onto the copper surface and

the cells can be varied in both depth and size dependent on the specification required. The 'cell volume' is the measure of the ink capacity of the cell and this is determined by the width and depth of the cell. The depth and shape of the cells is a key factor in the efficient delivery of a uniform ink film to the printing plate and allows the operator to vary the volume of the ink film by changing the anilox roller to the achieve the correct color.

The number of cells on an anilox roller are measured in cells per linear inch (CPI) or the cells per centimeter (CPC). **(See Figure 5.6).** As cell count increases, ink delivered to the plate decreases, but as the line screen resolution of the image increases, the CPI should also increase.

After the roller has been engraved, a ceramic coating is applied to it giving it a very hard and

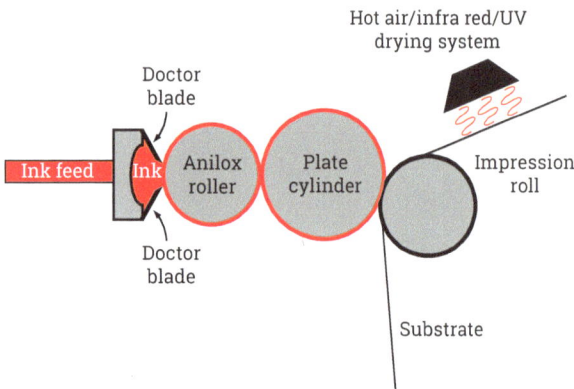

Figure 5.3 - The chambered doctor blade system. *Source: 4impression*

Figure 5.5 - Anilox roller. *Source: Mark Andy*

Figure 5.4 - Actual chambered doctor blade system. *Source Tresu*

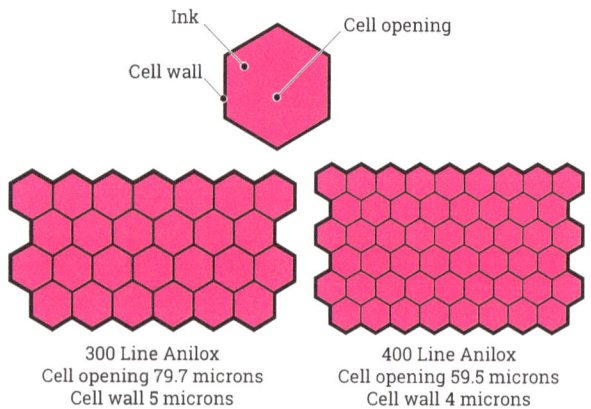

Figure 5.6 - Examples of anilox cell structures

durable surface finish and extending the life of the roller.

The screen angle of the anilox cell is very important. Print faults can occur that are caused by incorrect angles of the engraved cells. This can create a 'moiré' pattern which is a result of a screen clash between the anilox cell angle and the angle used for imaging the dot on the printing plate. Anilox cells are normally engraved at an angle of 45% - 60% from the center of the anilox roller in the horizontal position. **(See Figure 5.7).**

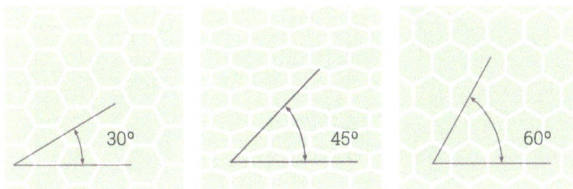

Figure 5.7 - Anilox roll cells are engraved at one of three angles: 30°, 45° or 60°.

Careful monitoring of the ink densities delivered by each anilox roller is important and is strongly recommended. Each roller should be inspected on a regular basis for any damage or reduction in the cell depth and a record of the ink volume of each anilox should be kept to ensure that the anilox specification is correct for the next print run.

ENGRAVING THE ANILOX ROLLER

The engraving of the cells on a modern anilox roller is now predominantly carried out using a laser engraving system. Previously the majority of anilox rollers would be engraved using mechanical engraving and this method of engraving is still in use today. The cell formation for this type of engraving differs in shape from the laser engraved cell.

Mechanical engraving can produce a Pyramid cell **(See Figure 5.8)** which is a fully inverted pyramid shape. The other mechanical cell is the quadrangular cell, **(See Figure 5.9)** which is a truncated pyramid shaped cell offering a higher ink release factor and good uniformity over the face of the roller.

Anilox rollers are specified by the number of cells per linear inch (line screen) and this can range from 250 to

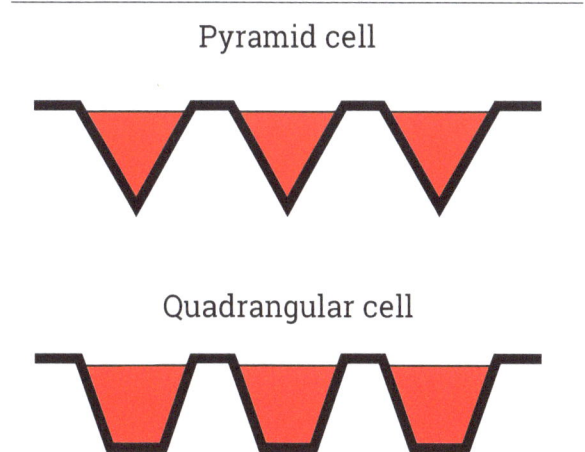

Figure 5.8 – 5.9

upwards of 2000 line screen. The majority of anilox rollers are screened at 800 LPI, but there has been an increase in the demand for 800-1200 LPI.

When preparing the next print run the press operator will select an anilox roll with a higher or lower LPI depending on the content of the graphics to be printed. If a heavy volume of ink is required the operator would select a low LPI screen value. A higher LPI value would be used for finer detail, for instance in producing CMYK fine screen printing.

Graphics can often be a combination of both fine screen and solid content and therefore it is necessary to have a portfolio of anilox rollers to suit each specific color/screen requirement.

LASER ENGRAVED ANILOX ROLLERS

Modern anilox rollers are laser engraved. This method of engraving gives the printer a considerable range of very accurate and consistent LPI screen options. The laser cut cell differs from the mechanically engraved cell and comprises of a circular 'scoop' style cell shape. The size and depth of each cell can be controlled and reproduced to a very high degree of accuracy which means that a specific anilox value can be repeated as and when required. This is a big advantage when re-engraving worn anilox rollers.

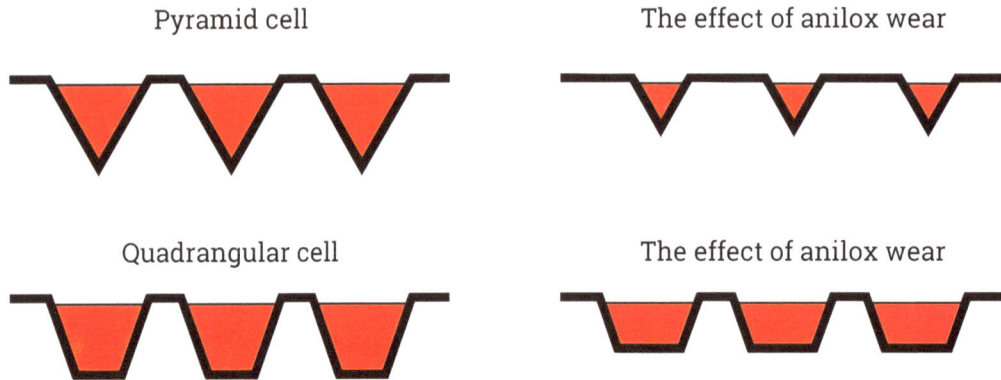

Pyramid cell

The effect of anilox wear

Quadrangular cell

The effect of anilox wear

Figure 5.10

WEAR ON THE ANILOX ROLLER

Anilox rollers are subjected to considerable abrasive wear. The abrasive action of some ink pigments and the impact and pressure of the doctor blade will reduced the cell depth and thereby affect the ink volume. **See Figure 5.10**.

ANILOX ROLLER MAINTENANCE

The maintenance of the anilox roller is most important. If any dried ink or varnish remains in the bottom of the anilox cells then the volume of ink delivered by the cell will be reduced and the color density will change. The cleaning, storage and identification of the anilox roller has to be methodically carried out and the operator needs to have easy access to each roller and know that it is suitable for reuse with a guarantee that it will deliver the same volume of ink as when previously used.

It is important that the anilox roller is thoroughly cleaned at the end of a print run, in particular any anilox rollers that have been used for solvent based or aqueous ink systems. These inks will dry rapidly if left to stand and cleaning must be done immediately to avoid ink drying into the anilox cells and creating a problem knows as plugging. This occurs when a blocked cell creates tiny pinholes in the printed image. Cleaning can be more effective if a fine brush is used to remove the ink, but the use of a brass brush is not recommended as this can damage the anilox surface.

The most effective method of cleaning the anilox is by using an ultra-sonic cleaning unit. **(See Figure 5.11)** This is a very effective and environmentally safe cleaning method. The ultrasonic cleaning system works through the turbulence that is created when air bubbles, which are generated by the system, implode on the surface of the roller and ultrasonic waves bombard the cells agitating the cleaning solution in the cell cavities and removing any dried ink. The process can be improved if the cleaning solution is heated to a recommended temperature.

Figure 5.11 - Ultrasonic anilox cleaning

THE FLEXO PRINTING PLATE

There are two types of plate material used for flexographic printing; rubber and photopolymer. Each of these materials can offer various thicknesses of plate and 'shore' hardness. The majority of label work printed by the flexo process is produced using polymer plates and therefore this Chapter will focus on the imaging and mounting of the polymer plate.

The structure of the polymer plate comprises of a

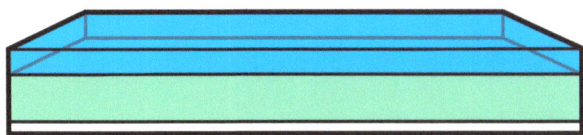

Figure 5.12 - *Flexo plate structure prior to exposure. Source: 4impression*

filmic base giving the plate its stability. The plate is coated with light sensitive polymer, which varies in depth dependant on the thickness of the plate required. **(See Figure 5.12).**

IMAGING THE PLATE

There are two methods used for imaging the flexographic plate.

1. Contact imaging, where the imaged film is placed in direct contact with the plate being imaged and then exposed to ultra violet light which establishes the image on the plate.
2. CtP, computer-to-plate where the image is created directly on to the polymer plate using a laser controlled by a digital file.

CONTACT IMAGING

CtP imaging is very much the dominant system used in the label industry, but film based contact imaging is still in use today.

The film, which can be negative or positive, is created from a digital file placed in direct contact with the polymer plate and then exposed to a UV light source. The three main processes involved in contact

imaging are as follows:

1. Imaging the flexo plate - the plate is back-exposed (UV) to create the base of the plate and then front exposed (UV) through the negative to produce the image and also harden the polymer image.
2. The plate is washed out in a wash-out unit to remove the unexposed photopolymer
3. The plate is dried and then is post-exposed and finished.

PRE-EXPOSURE

This is the process of exposing the back of the polymer plate which is to be imaged, to a UV light source. This exposure creates a limited polymerisation at the base of the plate and thereby gives increased stability to the plate and aids the anchoring of the dot to be printed.

MAIN EXPOSURE

The main exposure is made to the face of the polymer material, through the film and establishes the image that will appear on the finished plate. **(See Figure 5.13).** If using a negative film, the transparent areas of the film allows the UV light to pass through onto the polymer material and the polymerisation process hardens the image areas thereby creating the detailed solid areas and halftone dots.

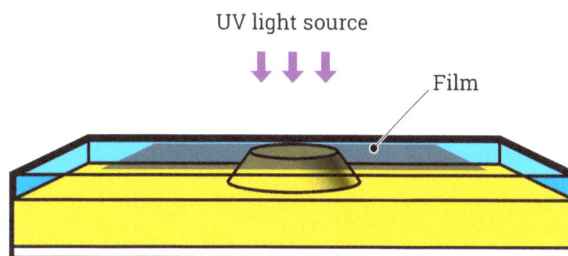

Figure 5.13 - *Exposure of flexo plate. Source: 4impression*

The quality of the imaged plate is dependent on the quality of the film used, correct exposure times and the condition of the processing equipment.

The exposure time is influenced by the condition of the exposure unit. It is important that regular testing procedures are carried out to ensure that the UV light source, generated by the UV tubes are transmitting at the specified levels and that the vacuum system which secures the film and polymer material together is operating correctly, providing a 100% contact of film to plate.

Additional tests should be made to identify any variation in the polymer material that can occur from one batch of plate material to another.

WASH-OUT

The areas of the polymer which have not been exposed to the UV light and therefore not been polymerised can now be removed. This process is called the wash-out process.

The depth of the relief image area will vary with the thickness of polymer being used and also the amount of back exposure the plate received prior to the imaging process.

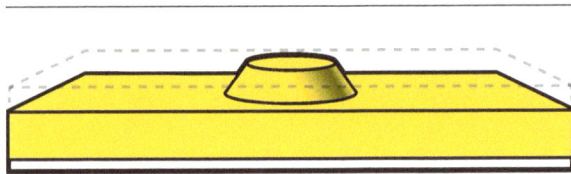

Figure 5.14 - Wash out of flexo plate

The wash-out of the plate is carried out in a washout unit. The plate is firstly positioned in the unit and made secure whilst a rotating brush is placed over the plate and set in motion to gently brush away the non-image areas. **(See Figure 5.14).** At the same time a solution of water and soap (approx 40 degrees centigrade) flushes away the waste polymer. Filtering of the waste is also done to ensure that disposal of the waste meets any environmental requirements. This wash-out process can also be done using solvent, but care must be taken to ensure that the wash-out time is restricted to a minimum, to avoid any swelling of the polymer material, as this will affect the dot formation (in particular the fine tones as the dots will become distorted).

DRYING

After the imaging and wash-out process is done, the drying of the plate is completed in a heated air cabinet. The amount of drying time required to ensure that the plate is fully dried will vary dependent on the thickness of the plate material and the type of wash used. Any solvent which remains in the plate has to be removed using the drying process. It is important that the specified drying time and temperatures used are maintained and that the plate is evenly dried across the entire plate surface. It is also advisable to allow the plate further drying time at room temperature to ensure the correct finish.

POST-EXPOSURE

Post exposure is an additional exposure to a UV light source (without using the film) and allows the parts of the relief image area that are not fully polymerised to be fully cross-linked (hardened) thereby ensuring that the finished plate is even over the whole surface.

COMPUTER TO PLATE - DIRECT IMAGING

Flexo plate imaging can also be done without the need for film originals, using a process called CtP or (Computer to Plate).

Digitally driven 'direct imaging' systems are now used extensively giving the label printer fast, accurate and consistent screening and plate imaging.

The file which contains the image to be printed is transferred to a device called an imager and the image is created using laser imaging. This eliminates the need for film. **(See Figure 5.15).** Direct imaging of the flexographic plate uses a high powered laser to

Conventional

| File | ▶ | Film | ▶ | Plate | ▶ | Press |

CtP

| File | ▶ | Plate | ▶ | Press |

Figure 5.15 - Plate imaging - convertional versus CtP

burn the image 'directly' on to the polymer plate, in a single operation. After the imaging process the plate is washed out to remove the surplus 'non-image' area of the plate which is then dried and finished. This method of imaging the plate reduces the costs which are associated with the multiple processes needed when using the film contact system. In can be seen in **Figure 5.16** that the workflow for contact imaging has eight stages compared to direct imaging which has only four.

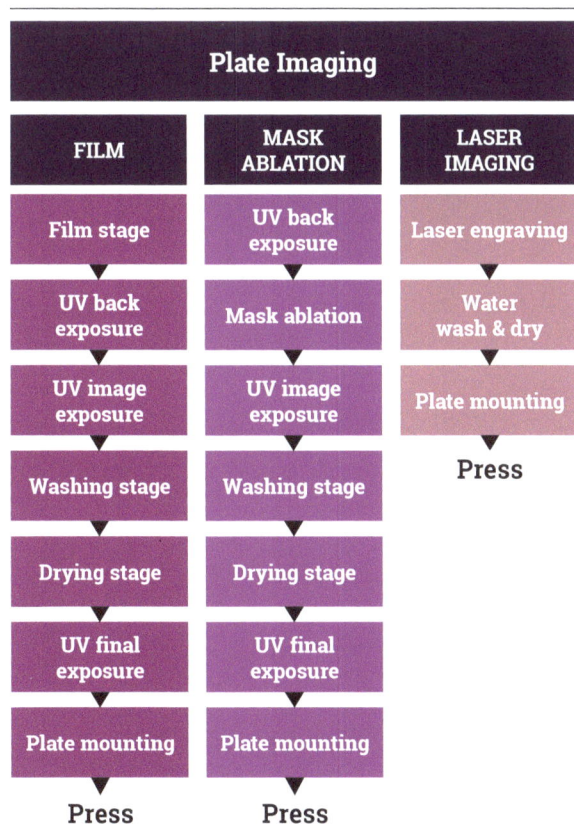

Figure 5.16 - Comparision sequence - film, mask abation, direct laser imaging processes

ABALATION IMAGING

An alternative laser imaging system is the laser ablation system in which an infrared-ray ablation layer (black mask) is coated onto the surface of the polymer plate and covers the entire plate. The plate is then imaged by removing the areas of the mask which form the image. This removal uses a high-powered digitally driven infrared laser to remove the masked areas that will form the image, revealing the polymer underneath which is unaffected by the laser ablation. A UV light source is then applied which polymerizes the image area where the mask has been removed.

A second exposure from the back of the plate generates the base of the plate. The plate is then washed out to remove the non-polymerised area - including the mask. The plate is then dried and given a final UV expose to fully harden the polymer.

PLATE MOUNTING

Preparation is the key word when mounting flexo plates. The print cylinder surfaces should be thoroughly cleaned to ensure that the surface of the cylinder is not contaminated with blade cuts, grease or debris. Any contamination of the cylinder surface will cause problems with the print quality. Any foreign particles trapped between the surfaces of the print cylinder, the mounting tape and the back of the printing plate or any greasy residue will affect the adhesion power of the mounting tape causing plate lift during the print run.

Double sided filmic adhesive tape is used to fix the printing plate to the printing cylinder. These tapes can be varied in thickness to allow for any variations in the diameter of the print cylinder and can be used to compensate for any under or oversized print cylinders. It is critical that the outside peripheral of each printing plate is exactly the same. If this is not done, accurate print to print register will be impossible.

It is strongly recommended that the same brand of tape should be used on each set of print cylinders. Some tape manufacturers incorporate a thin layer of foam within the mounting tape which assists in smoothing out any small deviations in the plate cylinder.

MOUNTING FLEXIBLE PLATES

Even though plate mounting is usually done using a plate mounting system equipped with mechanical and optical aids, the mounting of a 'flexible' printing plate **(See Figure 5.17)** is dependent on manual skills.

Figure 5.17 - *Plate mounting unit. Source: Nuova Gidue*

The one rule that must be followed is to ensure that the plate will be in the correct position at the first attempt of mounting. This involves a careful check to ensure that the registration lines on the leading edge of the plate are correctly in-line with either the optical system on the mounting system or alternatively the engraved grid on the print cylinder. These engraved lines run both horizontally and circumferentially round the print cylinder. **(See Figure 5.18).**

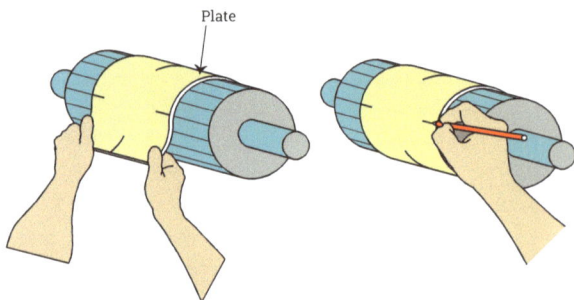

Figure 5.18 - Mounting plates by hand using guidelines

Generally speaking the plate is usually mounted in the center of the cylinder and it is recommended that the center of the plate is marked and lined up with the mark located with the center of the cylinder.

It is important that a little extra time is spent in ensuring that the leading edge of the plate is in the correct position before completing the full mounting as any removal of the flexible plate to 'try again', can distort and stretch the plate making any attempts at re-mounting very difficult.

The bearer bars are strips of plate material which are positioned on the outside area of the printing plate **(See Figure 5.19)** creating an even impression over the plate cylinder. This reduces the plate bounce which can occur between the gaps in the image.

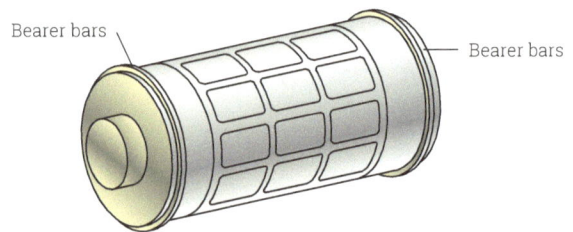

Figure 5.19 - Flexo printing unit showing position of bearer bars. *Source: MPS*

THE FLEXO PRINT CYLINDER
Regular maintenance of the press is very important. Bearings, print cylinders, gears etc. are subject to wear and it is advisable that replaceable parts are held available to avoid any delays in the event of a breakdown. Presses are subject to stresses and strains caused by the use of heavy tooling units, continuous running and poor maintenance. Cylinder bearings have a limited life and a procedure for the regular checking of the print cylinders should be carried out including cleaning, checks for wear and adequate lubrication.

With the flexographic, letterpress and litho processes the printing plate is located on the print cylinders. The cylinder needs to have accurate and even contact with the inking rollers and the surface of the substrate or in the case of the litho process the offset blanket. The plate/print cylinders should run perfectly true with an accuracy of ±0.025 mm ensuring that the pressure on the adjacent rollers is consistent. Accurate measurement should be made when the press has been running for a short period to allow the running parts to warm up.

PRINT CYLINDER GEARS

The print cylinder gear fits onto the end of the plate cylinder. **(See Figure 5.20).** These gears are manufactured to be meshed at a certain depth with the gear it is driving. This is called the pitch diameter. When meshing the print cylinder gear with the impression roll gear this pitch diameter is critical to accurate register. Too deep or too shallow mesh will cause print registration problems between one color and the next. Regular lubrication is preferred to intermittent applications as this will maintain a constant film of oil and even out temperature fluctuations.

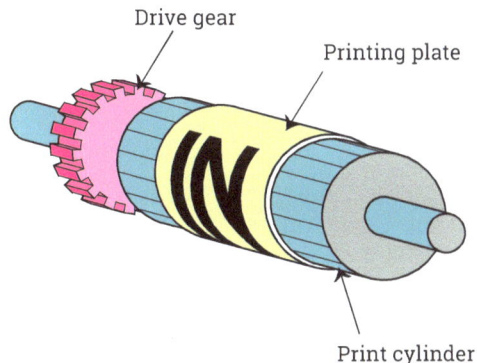

Figure 5.20 - Location of drive gear on the print cylinder

The circumference of the print cylinder is based on the repeat length of the printed image and is calculated in inches, millimeters, or the number and size of gear teeth; allowance has to be made for the 'thickness' of the printing plate and mounting tape.

The formula for this calculation is as follows;

A print repeat length of 12 inches (or 96 1/8' gear teeth) equals a circumference of 300.8 mm, the print cylinder must have a circumference that is smaller by 3.14 (Pi), to allow for two thicknesses of plate and mounting tape, 3.14 x 2 (1.7 + 0.3) mm and a small allowance for the effects of the printing pressure and thermal expansion, say 0.01 to 0.03 mm.

SERVO DRIVE TECHNOLOGY

Servo driven motor technology is now widely used in the commercial, packaging and label industries. Each print and impression cylinder and anilox roller has its own servo motor. **(See Figure 5.21)**, including the web transportation and web tension rollers and if required, the chilled rollers as well.

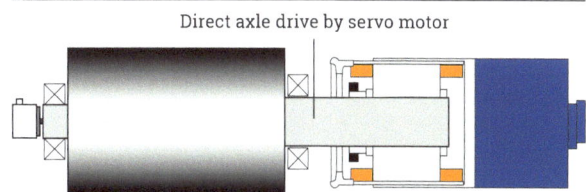

Direct axle drive by servo motor

Figure 5.21 - Direct drive servo system. *Source: Gallus Ferd, Rüesch*

The servo motors are connected to a motion control unit via fibre optic cables. This unit controls the synchronising and function of the motors used on the press and commands each drive independently or as a whole system.

Servo or shaftless drive systems have re-invented the mechanical drive, removing the need for any mechanical drives or gear train and therefore many of the print problems associated with gearing.

Servo systems have substantially reduced substrate waste, speeded the makeready times and improved print registration and print quality.

Changeover from the finished job to the next job (called the makeready time) can be considerably reduced using servo systems, as it gives the facility to save job settings (job storage) and to recall the data for pre-register, auto-registration and print length when the job is repeated.

SERVO DRIVES - ADVANTAGES
- No motors, gearboxes, clutch or gear trains required
- No gear marking
- No requirement for special gearing
- Infinitely variable print length
- No gearing issues between the plate and impression cylinders

	Mechanical Solution	Shaftless Solution
Synchronisation	Gears: Gears have inherent errors from the machining process. Mechanics wear out with use.	Cylinders driven from servomotor provide better synchronisation and eliminates mechanical wear
Timing	Mechanical device or sensor requires the press to move and create waste before timing is accomplished.	Servomotors can be electronically adjusted prior to machine movement.
Repeat size change	Limited to gear pitch of cylinders. This creates waste or design limitations.	Electronic gearing provides infinitely variable repeats.
Speed	Gears limit speed and cause vibration from resonant frequencies.	Gears and associated limitations are removed.

Figure 5.22 - A summary comparison between mechanical and servo systems.

- Quicker job changeovers with faster make-readies and less waste
- Automatic registration
- Computer-controlled simplified operation
- Lower maintenance costs, greater press life and fewer moving parts

PLATE CYLINDERS AND REPEAT LENGTH

Flexographic presses used in the label industry are able to operate with a 'variable' repeat length facility. As the plate cylinders are removable and not in a fixed position, printing from cylinders of different diameters can be used, thereby allowing the printer to minimise the amount of material between each label gap and accommodate the varying print lengths required when there is a change in the label dimensions..

Step motors provide an independent drive to each roller and this allows the speed of the plate cylinder to be varied without the need for a plate gear change.

INKS AND DRYING SYSTEMS

There are three main flexographic ink systems used in the label industry. These are water-based, solvent-based and ultraviolet (UV). There are two lesser used

ink systems known as electron beam (EB) cured inks and two-part chemically-cured epoxy inks which are used to obtain very high product resistant coatings.

In this chapter we deal with the three main ink systems used for label printing and the types of drying/curing used.

All the inks used in the flexo, gravure and screen processes are liquid inks which are dried or cured after each individual color is printed (ie multiple colors are printed wet ink onto dry ink).

Solvent based inks are not widely used in the label industry. Previously water-based inks provided the bulk of label work produce by the flexo process, but this has now changed and UV is the dominant ink system used in label manufacturing.

Printing inks consist of four component parts:

- Colorants (pigment or dye)
- Vehicle (the binder)
- Additives (reducers, waxes, driers, flow agents)
- Carrier (aqueous or solvented).

The colorant provides the color to be printed and can be pigmented or dye based. The density of color is dependent on the volume of pigment immersed in the vehicle and the amount of pigment is known as the solid content. Generally the more solids there are in the ink the more difficult it is to transfer the ink to the substrate. As flexo inks are 'thin' with a low viscosity it can sometimes be difficult to achieve the correct density of color required. This however does not apply to UV based inks which are 100 % solids.

The vehicle consists of resins in which the pigments or dyes are mixed along with the additives and the carriers (i.e solvent or water) and it is this mixture of resins, pigments and additives that determine the characteristics and performance of the ink.

SOLVENT BASED INKS

Solvent based inks use an organic volatile solvent, mainly ethyl acetate and alcohol, used as a blend and not a single component. This ink system is also used for the printing of self-adhesive labels by the flexo, gravure, screen and inkjet printing processes.

The solvent controls the viscosity of the ink and is excellent when rapid drying is required. The viscosity of the ink is controlled by the amount of solvent in the ink. If solvent is introduced into the ink the viscosity reading will be lower because the ink will be thinner. If however the ink is allowed to thicken through the lack of solvent, then the viscosity reading will be higher.

Ink viscosities are very important as the color being printed will be adversely affected if the readings fluctuate. Viscosity readings can be done manually using a 'zhan cup'* or if the press is using a circulatory ink pumping system then these readings can be automatically carried out and are usually linked to a system that automatically introduces the solvent as required. Drying is done through hot air driers which drive off the solvents as the substrate passes through the drying hood. The extracted air can then be passed through a solvent recovery system.

Press wash-up and clean down between each run is done using solvents that match the ink solvent system being used.

*Zhan cup - definition

Dip calibrated viscosity measuring device. A stainless steel cup with a tiny hole drilled in the center of the

bottom of the cup. After lifting the cup out of the ink, the user measures the time until it stops flowing to assess its viscosity.

DRYING SYSTEMS: INFRA-RED/HOT AIR

Infra-red is the method of accelerated drying which uses either hot air or direct heat or a combination of both. **(See Figure 5.23).**

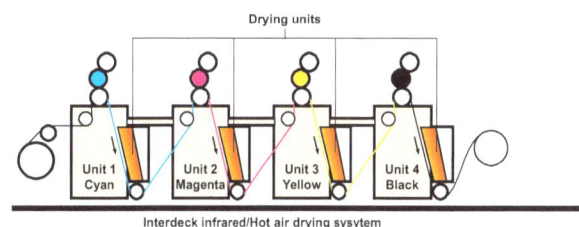

Figure 5.23 - Drying system for water-based/solvented flexo

For flexo presses which are printing with water-based or solvented inks the drying system is infra-red or hot air or a combination of both. This type of drying system is compatible with many of the web-fed presses used in the self-adhesive industry.

Infra-red drying is not suitable for filmic materials as the heat that is generated creates problems with shrinkage and stretching of the filmic substrate, in particular lightweight unsupported film.

WATER-BASED (AQUEOUS) INKS

Water-based inks, also known as aqueous ink, were the original inks used on the very earliest flexo presses. These early inks were dye based aniline inks.

The modern water-based ink system is capable of producing very high quality printing and is used extensively within the label industry.

This type of inking system is more environmentally friendly than solvent based inks, particularly as they do not present the fire risk posed by solvented systems.

The range of colors that are available are the same as both solvent and UV systems and these inks will print on paper, metallised substrates and give excellent results on non-absorbent plastics/film i.e. polyethylene, polypropylene and PVC. It is recommended that an in-line corona discharge pre-treatment is given to non-absorbent substrates to assist the ink key to the substrate and to improve the ink lay down.

Water-based inks do contain a small percentage of solvents and chemicals, but this is kept to a minimum. These solvents can affect the viscosity of the ink and color should therefore be carefully checked during the print run.

ULTRA VIOLET - UV FLEXO

UV flexo is a print process in its own right and not just a variation on the conventional flexo process. The main principles remain the same, but the facility to print a much finer dot, with a thicker ink consistency and 100% solids content is similar to the inks used for letterpress and the offset process, but with a lower viscosity.

The introduction of UV flexo has led to the development of a range of anilox rollers that give the printer a number of options, particularly when printing very fine tone work. It has also allowed the use of file origination that is the equivalent to that used in the litho process. This has made a big impact on the manufacture of self-adhesive labels and allows the printing and converting of filmic and metallic substrates, using an oil based UV system. The drying or curing of UV ink is through the reaction of the ink chemistry to an ultraviolet light source. The curing process is a photochemical reaction that occurs when monomers and oligomers are mixed with photo initiators and then exposed to a UV light source. The curing process takes 1 - 2 seconds.

UV inks are a 100% solids system and do not contain solvents as the main carrier. This means that they are not subject to fluctuations in the ink viscosity and can overcome any color variation during the print run.

Because UV curing is a rapid process the press can be run at high production speeds, but curing can be affected by the color and density of the ink film and the intensity of the UV light source. Opaque inks are the most likely to have a curing problem as the UV light source must penetrate fully through the ink layer to ensure a full cure.

UV ink does not dry unless exposed to ultra violet light and this means that any press stoppage does not create the problem of rapid ink dying as experienced with water-based and solvented inks. Washups are quicker and end of shift washups are not necessary. Also there is not the problem of dried ink in the anilox cells known as plugging.

Advances in polymer plate and imaging technology have further advanced the quality of UV flexo printing and the process can now hold a dot formation of 1%, allowing very fine tone work to be produced.

The polymer used for UV flexo plates is thinner and harder, similar to that used for letterpress polymer plates. This allows the dot produced to be more stable and therefore not subject to flexing during the printing operation. It also facilitates the accurate transfer of ink film which in turn means that dot gain is reduced and the 'squashed halo' effect which is inherent in conventional water-based and solvented flexo is eliminated.

UV CURING SYSTEMS

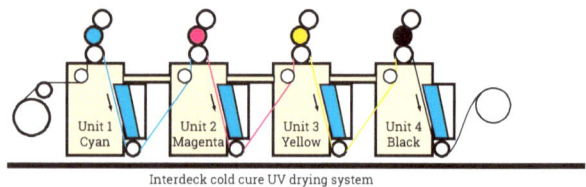

Interdeck cold cure UV drying system

Figure 5.24 - Curing system for UV flexo

The modern UV system is far easier to operate and maintain than the earlier systems. The power of the ultra violet intensity can be adjusted to suit a particular ink curing requirement and the controlled removal of the unwanted heat generated by the infra-red wavelength ensures that the web is only exposed to a minimum amount of heat. Heat is removed by the use of cooled air which is blown across the lamps and then extracted by the use of water cooled plates which also reduces the amount of heat generated by the lamps. These methods of heat removal allow a UV ink cure that is 'cold cure' and makes the printing of light filmic substrates much easier.

Figure 5.24 shows the position of the UV drying / curing units in the printing press. The curing systems used can be run at very high production rates as the curing of the UV ink takes place rapidly. However drying speed can be affected by the color and density of the ink film and the intensity of the UV light source.

LED UV CURING SYSTEMS

The term LED stands for a light emitting diode. The UV light source is generated by passing an electric current across a series of diodes which then generate high intensity UV energy. This technology is incorporated into a unit suitable for UV curing on the press. LED lamps are smaller than mercury units **(See Figure 5.25)** which makes them ideal for the narrow-web presses used in the label industry. The lamps come to full brightness without any warm-up time and have a much longer lifespan than the standard air-cooled mercury UV arc-lamp system used on label presses. LED systems can reduce production

Figure 5.25 - LED Curing

downtime and remove the problems that can occur when the lamp shutter mechanisms fail. LED systems do not require the same routine checks to monitor the energy densities of individual lamps as are required

for mercury lamps to ensure there is no degradation in output.

Some of the advantages of LED systems are summarised below;

- Low power lamps are cooler with less heat radiated onto the substrate
- Approximately 50% reduction in energy costs
- More effective curing of opaque inks
- Reduced maintenance on shutter system
- No energy degradation – longer lamp life
- No generated ozone issues.

TONAL RANGE (AQUEOUS/SOLVENTED)

What are tonal values? All printed graphics are made up of a printed dot formation, the lighter the dot area, the lighter the color, the more dense the dot formation the darker the color. **(See Figure 5.26).** This variation of dot density and size gives the various tonal values required to form the printed tones.

Dots break up at 1-2%

100% dot   50% dot

Figure 5.26 - UV flexo dot break up at 1-2%

IDENTIFYING THE FLEXO PROCESS PRINTED WITH SOLVENTED OR WATER-BASED INKS

It is necessary to use a magnifying glass to identify the method of printing which has been used on a particular graphic reproduction.

Each different printing process has a particular characteristic which can be easily identified. As explained the printed image is made up of differing densities of dot and there are limitations on the size of the dot which can be printed by some of the print processes. Generally speaking the size of the dot which can be produced by the traditional solvented and water-based press is between 3-4%.

The modern polymer plate used for the flexo process can hold a one to two percent highlight dot. By looking at the highlight areas the printer can identify the size of the dot and also the quality of the print.

Flexo plates are a relatively soft plate compared to the much harder letterpress plate and this soft construction can affect the print quality. The softer the plate the more likelihood of a squashed effect on the printed dot. However, softer plates will transfer the ink film more smoothly than harder plates. This squashed dot can easily be identified by a halo effect on the printed dot.

This poor quality printing is the result of a combination of a soft printing plate, too much impression and a low viscosity ink.

One of the biggest problems confronting the printer is the problem of dot gain. This effect is created by an 'increase' in the specified printed dot size which affects both the tonal values and therefore the color being printed.

Dot gain can be a result of incorrect plate imaging but is generally a result of incorrect impression settings or poor engineering or wear on the press and the tooling being used. Dot gain can also increase if a harder 'non- cushioned' plate mounting tape is used. Alternatively a softer 'cushioned' tape can reduce the gain but does not always provide for a good ink transfer for the solid printing areas and can also create pinholes in the print.

IDENTIFYING DOT GAIN
To effectively measure dot gain it is strongly recommended that the dot gain characteristics of each individual press is first established. This is not a difficult operation and requires a printed sample to be taken on the press being used and using a printing plate with a known dot specification. Press settings must be carried out to get the optimum print quality. After the print sample has been produced the gain can be accurately measured to establish the dot gain of the press.

IDENTIFYING UV FLEXO
The modern flexo plate used for UV flexo tends to be harder and when used in combination with a much higher viscosity UV ink, gives the high quality flexo printing now produced in the label industry.

The UV process does not experience the problem of thin inks, and dot/print squash is very much reduced .The modern UV flexo press will print a very crisp dot of 1-2%.

FLEXO ON THE PLATFORM PRESS
The use of the flexo process on combination presses is now commonplace. The process is compatible with all the major printing processes and often forms the main process of the combination press.

ADVANTAGES AND DISADVANTAGES OF THE FLEXO PRINT PROCESS

Advantages
- Suitable for high speed web-fed printing.
- Suitable for most materials.
- Ideal for food packaging/newspapers.
- Ideal for non-absorbent stocks.
- Wide or narrow web
- Suitable for combination printing.
- The pre-press costs are significantly less than for gravure.
- The printing unit design allows for the repeat length to be easily changed.

Disadvantages
- Limited halftone reproduction
- Process is subject to bar marking and ghosting.
- Extended wash-up
- Rapid drying ink creates press difficulties
- The ink system can use relatively volatile liquids

FLEXO – SUMMARY
Flexo is far and away the predominant label converting method both in North America and also in Europe.

Advances in all of the components of the process inks, plates, anilox rolls, and the presses themselves have all played a significant role in the evolution of the flexo process, and they continue to do so today.

INKS

Developments in inks formulations with increased color strength, able to deliver the required film thickness and deliver high resolution images, have had a significant impact on the development of the process.

A significant trend has been the transition from solvent based inks to water and UV based inks.

Environmentally friendly water-based ink, many formulated with renewable plant based resins are on the increase, as are low migration UV flexo inks for use on primary food packaging.

The key benefits with UV flexo include more brilliant colors, zero or very low VOCs and the reduction of clean ups between shifts.

ANILOX DEVELOPMENT

Directly related to the trends and advances in flexo inks is the method used to deliver them onto the substrate – the anilox roll.

The introduction of ceramic laser engraved anilox rolls has enhanced ink distribution and enabled an increase in cell concentration. The practice of engraving deeper and at higher line counts has contributed to both color and image quality.

The innovation of fibre optic laser technology that is able to create a cell profile with much smoother cell wall linings, thereby allowing a better ink release, has been a key factor contributing to image quality

FLEXO PLATES AND WORKFLOW AUTOMATION

Flexo plate improvements continue to play a major role in print quality. The following factors have been important:

- CDI (Computer Direct Imaging). CDI imaging heightens results to levels that previously were not available. The use of new screening algorithms achieves a very wide tonal range with smooth and detailed highlights and graduations going to zero.
- CtP in flexo platemaking computer to plate (CtP) and the use of laser ablation continues to move forward

WORKFLOW AUTOMATION

Automating the workflow is a major trend in flexo. Software workflows have eliminated several steps from the process resulting in huge timesaving and in error reduction.

SERVO TECHNOLOGY

Servo technology, which eliminates the traditional line shaft/gear box drive systems, has lifted the bar in press construction and created new opportunities for in-line flexo

Each plate, anilox and impression cylinder is able to receive its own servomotor along with web transport rolls such as tension rollers and chill stands, thus helping to eliminate inconsistency, whilst increasing flexibility and quality.

Registration control is vastly improved through the use of servos, thereby leading to a significant reduction in waste.

CONCLUSION

Today, flexo has carved out its own niche in the packaging and label printing sectors enabling cost effective, high quality in-line printing and converting for everything from short-runs to high-speed long runs on paper, film or board.

Flexo is perhaps the most versatile printing process in the world and the process is perfectly aligned with the future of a dynamic label and packaging industry.

Chapter 6

The screen printing process

Screen printing, also known as silkscreen printing, is one of the oldest methods of printing. It first appeared in China during the Song Dynasty (960–1279 AD) before emerging in Japan and a number of other Asian countries. Early silkscreen printing used silk mesh but as this material was not readily available in Europe the screen process was not introduced there until the latter part of the 18th century.

SCREEN PRINTING IN LABEL MANUFACTURING

The ability of the screen process to print high volumes of ink onto the substrate was a significant factor behind its introduction into the label industry. The opportunity to produce a clear filmic self-adhesive label that matched the graphics produced on bottles that were 'direct' printed opened up new opportunities for the self-adhesive industry particularly in the personal care, wine and spirits and pharmaceutical markets. The facility given by the self-adhesive labeling system to apply labels when required at the same time as the filling process, removed the need to 'direct' print empty bottles. The opportunity to eliminate the expensive logistics associated with the storage of pre-decorated containers was compelling. Screen printing is able to deliver a very dense white which can both mask product show-through and also provide a base that can be subsequently overprinted with other printing processes. This characteristic vastly increased the graphic options available for product decoration i.e. very dense color linked to fine tone work.

The development of the rotary screen process was the biggest factor in bringing screen printing into

the label sector and this move was further accelerated with the introduction of steel screen mesh and its use in combination with other UV ink systems.

TYPES OF SCREEN PRINTING LABEL PRESSES

The two types of screen press used in the self-adhesive industry are flatbed and rotary screen configurations. The majority of these press types are web-fed and can be operated as a dedicated screen press or as a combination of screen and other printing processes. The early Gallus T180 (UV curing) is a good example of the flatbed screen system whilst the later Gallus R200 (UV curing) is a good example of a rotary screen press. Both these presses used both the letterpress and the flexo process in combination with screen.

PRINCIPLE OF THE SCREEN PROCESS

Unlike the letterpress and flexographic printing processes which transfer ink from a relief plate, in the screen process the ink is pushed through a screen mesh onto the substrate being printed. **(See Figure 6.1).**

With the flatbed system the screen material, which

can be polyester or nylon, is stretched over a flat frame.

The screen is now ready for printing and can be placed in the press. Ink is poured onto the flat screen and with the substrate laying a short distance beneath the screen (this is called the snap off distance). A rubber squeegee blade then moves across the stationary screen and the ink is forced through the open mesh area (the image area) onto the substrate creating the printed image **(See Figure 6.2).**

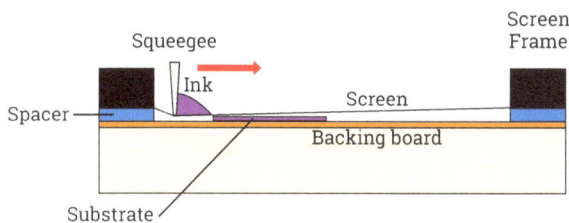

Figure 6.1 - Principles of screen process (4impression)

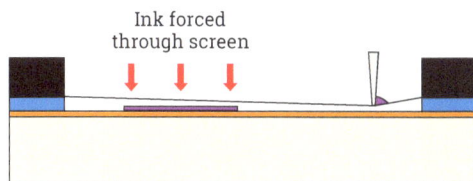

Figure 6.2 - Principles of screen process (4impression)

FLATBED SCREEN PRINTING

Compared to rotary screen, flatbed screen is a much slower process, because the substrate and the flat screen have to be stationary at the printing stroke. The flatbed system is a stop-go operation and therefore much slower than the rotary system. **(See Figure 6.3).** The flatbed screen was the first to be used by the label printer and though it has now been largely overtaken by rotary screen; it is still used

today.

The three important areas that control the quality of the screen printed image are the size and depth of the mesh material, **(See Figure 6.4)** the viscosity of the ink and the 'snap-off' distance between the screen and the substrate.

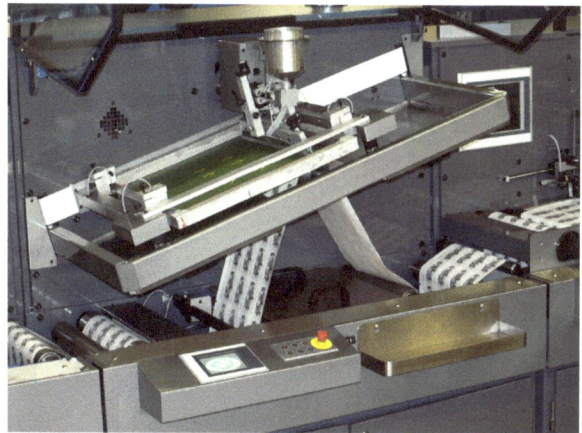

Figure 6.3 - Actual flatbed screen label press. *Source: Smag*

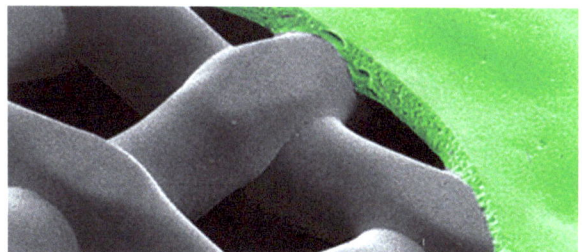

Figure 6.4 - Magnified exposed and washed out screen showing image and non-image area. *Source: Gallus Ferd. Rüesch*

In flatbed screen the 'snap-off' distance is most important and if it is not correct will create inferior print quality. The 'snap-off' distance is the gap between the mesh and the substrate and this determines the pressure required on the mesh by the squeegee blade. It is the squeegee blade which forces ink through the screen mesh, bringing the ink, screen and substrate into contact to produce the image. If the 'snap-off' distance is too great the image may become blurred, if the 'snap-off' gap is too little then smudging of the image will occur. Synthetic screen materials are now widely used in the manufacture of flatbed screen printing, with polyester being one of the most popular. These synthetic materials offer the flatbed printer a number of different mesh size options.

IMAGING THE FLATBED SCREEN

An overall photosensitive polymer emulsion coating is applied to the screen material and then dried before a positive imaged film is placed in contact with the flat screen. The screen is then exposed to a UV light source which hardens the emulsion in the 'non-image' areas thus making it insoluble in water. The emulsion in the 'image area ' remains soft and the screen is then pressure washed to remove the emulsion from the 'image' areas before the screen is then dried. **(See Figure 6.5).**

THE SQUEEGEE

The function of the squeegee blade in screen printing is to force the ink film through the parts of the screen mesh that forms the printed image. In flatbed screen the squeegee moves across the screen spreading the ink evenly across the back of the screen and creating the image. On the return stroke the squeegee then lifts slightly and brings the ink back across the screen ready for the next print sequence. In rotary screen the squeegee is in a fixed position and remains in contact with the screen surface as the cylindrical screen rotates.

Print quality will be affected by the type of blade used and the printer will need to select the most suitable blade type. The edge of the blade which is in contact with the screen can be varied in hardness and profile shape and will vary dependent on the type of job being printed. Any damage to the squeegee will affect the print quality.

ROTARY SCREEN PRINT

The introduction of steel mesh for screen printing led to an important change within the label industry. It allowed the steel screen material to be formed into a cylindrical shape **(See Figure 6.6)** which meant that

Polyester screen is stretched tightly over frame of wood or metal

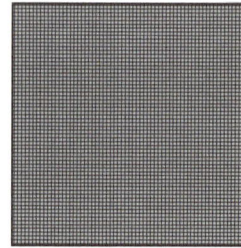

Screen before emulsion

Screen emulsion coated

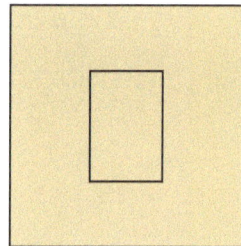

Image created photographically, emulsion in 'non-image' area is hardened

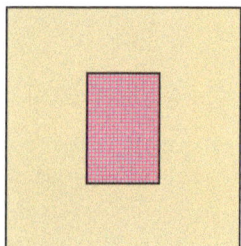

Emulsion in 'image area' is removed by pressure

Figure 6.5 - The stages of flatbed screen imaging

screens could be fitted into full rotary presses which are able to run at much higher speeds that the flatbed screen presses. Steel mesh screens can be produced with a wide range of screen value options allowing the printer to choose the most suitable mesh for each job. This choice provides more control over the volume of ink being printed and allows for very high coating weights of ink to be printed, way in excess of the other printing processes.

The rotary screen process adopts exactly the same

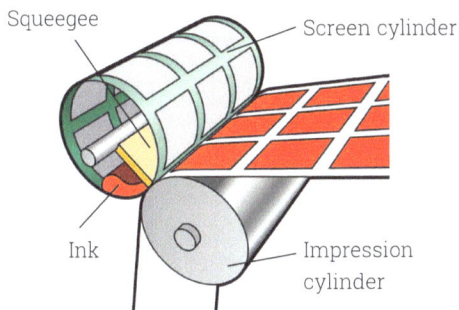

Figure 6.6 - Principle of rotary screen printing showing the ink and squeegee inside the screen cylinder. *Source: Gallus Ferd. Rüesch*

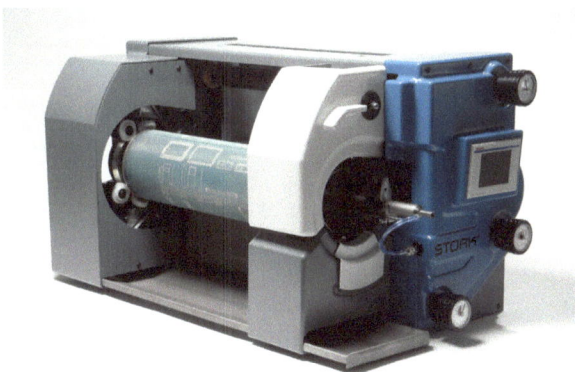

Figure 6.7 - Rotary screen unit. *Source: Stork*

principles as the flatbed system, but with some key differences. The imaged screen is formed into a cylindrical shape whilst the squeegee blade is placed into the screen cylinder in a fixed position. The screen

cylinder which then holds the liquid ink, rotates at the same speed as the web being printed and the ink is then forced through the imaged area of the rotary screen and onto the substrate **(See Figure 6.6 and 6.7).**

Unlike the flatbed system, which has no impression cylinder, the rotary screen prints against an impression cylinder, with the substrate running between the screen cylinder and the impression cylinder. This means that the two cylinders have a small point of contact which gives an excellent printed result. Ink levels are maintained within the rotary screen either manually or by the use of an ink pump.

ROTARY SCREEN IMAGING WITH FILM

The imaging of a rotary screen cylinder when using a positive film is very similar to the imaging of a flatbed screen. The rotary screens can be supplied to the printer already made up into the cylindrical shape (Stork system) or can be supplied as flat sheets that are then formed into the rotary cylinder by the printer. (Gallus 'Screeny' system).

With all rotary screen systems an end ring has to be fitted into each end of the screen cylinder. This gives the rotary screen the necessary stability and ensures that the screen rotates evenly during the printing operation.

The procedure for imaging rotary screens used in the label industry is as follows :-

1. Metal screen formed into cylinder and end rings fitted.
2. An overall photosensitive polymer emulsion coating is applied to the screen material and then dried.
3. The positive imaged film is accurately positioned in direct contact with the screen and then secured to allow the screen to spin in the exposure unit
4. The screen plus the secured film is placed into the exposure unit and exposed to a timed UV light source whilst the screen is rotating.
5. The emulsion in the 'non-image' area is hardened and becomes water resistant
6. The rotary screen is removed from the exposure unit, the film is removed and the

screen placed in the washout unit, in which the screen is pressure washed to remove the emulsion in the image areas.

7. The screen is removed from the washout unit and dried before making ready for the press.

ROTARY AND FLATBED SCREEN CTP IMAGING

CtP (Computer to Screen) imaging of both flatbed and rotary screens is now widely used. This method of imaging removes the need for film originals and eliminates the exposure process, power washing and drying of the screens.

The digital file which contains the image to be printed is transferred to the imaging unit. A high powered laser then 'burns' the emulsion away creating the image directly onto the screen. Laser engraving is a digital method of imaging both flatbed and rotary 'nickel' screens. It involves the removal of the emulsion coating in the image areas (i.e. the open areas of the screen). After this the screen requires no further processing and is ready for fitting into the press. In the case of rotary screens, the screen is imaged using a rotary CtP unit, whereas in flatbed imaging the screen remains flat.

CtP imaging reduces the costs associated with the multiple process operation needed when imaging by the traditional method of film contacting. A direct engraved screen produces excellent quality and consistency, with screen resolution of 2540 dpi being produced to allow fine line work with high contrast to be delivered.

Screens can typically be imaged in 15 - 20 minutes and because the lengthy drying process is eliminated productivity can be improved, giving a much faster 'turn round' compared to the conventional screen imaging method. Screen material is expensive but the ability to re-use and re-image screens, especially the rotary screens, has allowed some printers to include a facility which involves stripping off the unwanted image and recoating the screen.

SCREEN MESH MATERIALS

Screen printing materials fall into two categories fabric and metal.

Nylon and polyester are the standard mesh materials used today for flatbed screen printing, but screens can also be made from stainless steel wire mesh.

In both instances a single thread of material is woven together to form an accurate gauze structure.

Nylon is generally used when screen stability and tight print register are not required and because it is durable it is generally used for high volume single color work and for printing onto an uneven surface.

Polyester mesh has the same qualities as nylon, but it is more stable and is the material of choice when printing multiple-colors where fine registration is necessary.

Stainless steel wire mesh is more costly than nylon and polyester, but there are some advantages to be gained from using steel mesh screens.

Screens made of steel mesh offer greater screen stability and are even more durable than nylon and polyester. If there is a requirement for heavy ink deposits, steel screens would be the preferred mesh, but if required it will also produce fine print detail. Steel screens have a longer life particularly when printing with very abrasive inks.

MESH GRADE AND MESH/SCREEN COUNT

Screen mesh materials are identified by their mesh count and grade.

The term 'mesh count' is the number of threads woven into the mesh per centimeter. Screen materials are available in various mesh counts from 12–200 threads/cm.

The greater the number of mesh holes the better control the printer has of the ink film, particularly when the graphics require fine text and line work.

The lower the number of mesh holes the heavier the ink deposit, but this also means that the printer will struggle to print fine detail.

Mesh grading is the term used to indicate the thickness of the thread used. The mesh grade will affect the overall stability of the screen and also the depth of the screen mesh which controls the thickness of the ink being deposited.

The selection by the printer of the most suitable material and mesh count is most important. It will be influenced by the graphic requirement and will take into consideration the print detail, the stability of the screen and the effect on print registration and the thickness of the ink film required.

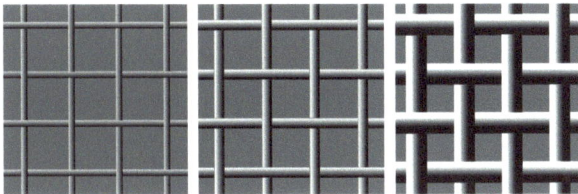

Figure 6.8 - Different grades of screen mesh

Figure 6.8 shows three grades of mesh assembled with the same thread count. Mesh thread thickness controls the size of the screen mesh open area. As the mesh count increases the print area of each cell is reduced giving greater control of the ink film.

ROTARY SCREEN MATERIAL

There are two main types of rotary screen materials used in the label industry. One is manufactured by SPG Stork and is called the 'RotoMesh' system and the other from Gallus is called the 'Screeny' system. There is a considerable difference in the structure of these two rotary screen systems. The 'RotoMesh' material is a nickel-based flexible sheet onto which a honeycomb shaped cell structure is engraved using

Mesh Count	75	125	215	305	305
Open Area %	40	15	25	13	11
Screen Thickness	150	100	80	80	80
Hole Diameter (microns)	214	79	59	30	28

Figure 6.9 - SPG Roto Mesh specification data - number of openings per linear inch

an electroforming procedure. This method of producing the mesh cell gives the printer an accurate and wide range of screen mesh options and print

possibilities **(See Figure 6.9)** for producing both fine tone work and heavy coatings of ink. 'Rotomesh' screens are very stable and can be de-imaged and re-imaged.

The Gallus 'Screeny' material is a microstructure formed from a stabilised fabric which has a photosensitive coating. The material has excellent structural properties which allow it to be formed into a cylindrical shape onto which end rings are fitted and glued. The forming of the rotary shape is carried out 'after' the screen has been imaged. This involves the welding or gluing of the two leading edges of the flat screen to form the cylindrical shape. Imaging is done with the screen laid flat onto an exposure unit fitted with a vacuum bed which can be contact imaged using film or direct imaged using CtP. Imaging time is typically 30 minutes.

The 'Screeny' system gives excellent print quality and is suitable for printing very fine line work and excellent solids.

COATING WEIGHT COMPARISONS

Figure 6.10 shows the 'coating weights' i.e. the thickness of the ink layer which can be achieved

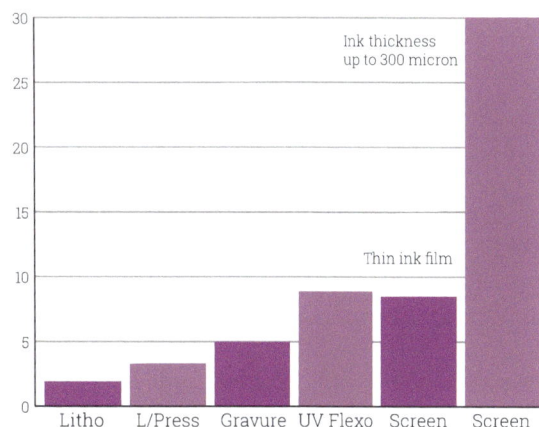

Figure 6.10 - Ink coating weight comparisons – screen versus other processes

using each of the major printing processes. It is clear that the screen process is able to deliver significantly more ink film than any other process. These figures are an approximation and it should be noted that developments in anilox roller and ink technology are likely to impact on these coating levels going forward.

SCREEN INKS

The majority of screen printing flatbed and rotary presses operating in the label industry use a UV ink system. Some dedicated label printing screen presses however do use solvented systems.

Screen inks used are of a relatively high viscosity. The ink must be viscous enough to avoid passing through the screen cells until the exact moment of the printing cycle when the ink is forced through the screen cell by the pressure of the squeegee blade. Inks with thixotropic properties (this is an ink that thickens up and affects the flow properties of the ink) can offer some advantage to the screen printer as it becomes fluid when agitation takes place or pressure is applied i.e. via the squeegee blade.

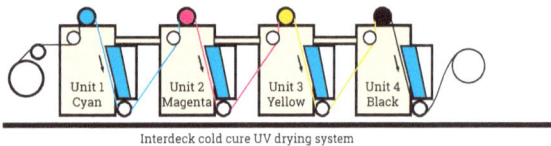

Interdeck cold cure UV drying system

Figure 6.11 - UV drying system for screen

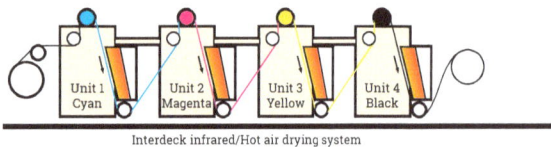

Interdeck infrared/Hot air drying system

Figure 6.12 - Infra red drying system for screen

Drying or curing when printing UV screen is done using the same web UV curing system as the litho, flexo and letterpress printing systems. **(See Figure 6.11)**.

Screen printing with solvented inks requires a closed system of infra red and or hot air drying and may also involve solvent reclamation. **(See Figure 6.12)**.

IDENTIFYING THE SCREEN PROCESS

The improvements which have been made in screen printing have made it more difficult to easily identify a screen printed image. Early screen printing could be identified by the 'saw edge' which appeared on the edges of the print.

Developments in screen materials now mean that finer woven meshes are available which allow much finer cells. Avoiding mounting the mesh at 90 degrees to the direction of the print has also helped improve the printing quality. Modern screen print can still be identified by the slightly uneven printed edge, but the viewer will now need a magnifying glass to correctly identify the screen process.

ADVANTAGES AND DISADVANTAGES OF THE SCREEN PROCESS

Advantages
- The main strengths of the screen process are:
- Prints very intense colors with excellent covering power
- Prints onto almost any substrate filmic, paper, metallic
- Prints onto curved, uneven and fragile surfaces
- Prints very thick, opaque inks suitable for tactile labels
- Prints unusual ink and coatings, micro encapsulation, metallic inks
- Prints brilliant colors, fluorescents, large particle size inks
- Controlled coating weights
- Solvented or UV drying
- Re-usable screens
- Fast makereadies
- Suitable for combination printing

Disadvantages
- The main process limitations are:
- Press running speeds tend to be lower, more so with flatbed
- Can be difficult to reproduce fine detail and small type

- Halftone reproduction requires coarser dpi rulings
- High costs of rotary screens
- Screen imaging can be a slow process

SCREEN PRINTING – SUMMARY

Screen printing in the narrow web segment has traditionally been viewed as a high-end add-on for those label converters pursuing the high added-value markets and who were willing to make significant investment in the technology.

Rotary screen printing is a very effective added-value printing process, being capable of printing thick ink layers with high accuracy at relatively fast speed.

Personal care packaging in particular has long used this process to achieve high opacity graphics, particularly on clear filmic materials (commonly referred to as 'no-label' look).

The key feature of rotary screen printing is the total flexibilty of the integrated units which allows for screen print to be added at any point during the print process and is perfect for integration into a combination printing press.

NEW APPLICATIONS

The screen process is increasingly being used for tactile and raised images, glossy and high lustre varnishes, metallic features and many other security features. Screens are also being used to apply pattern adhesives to a variety of converted products.

Two of the more interesting areas where screen has emerged are in the creation of braille information and in the printing of flexible electronics.

Braille images are typically printed with a UV varnish, at printing speeds of higher than 40 meters per minute (130 fpm), with thicknesses that can reach 300μ.

Rotary screen printing too has become the reliable and accepted printing solution for the application of electronic inks, facilitating high volume, low cost production. The printing of electronics such as RFID antennae on common substrates such as paper, film and textile using standard printing processes is rapidly growing.

TECHNOLOGY DEVELOPMENTS

Significant technology developments are making the screen process more accessible within the industry, and are reducing the cost of ownership.

The development of dedicated screen units for leading press manufacturers, for example, continues to facilitate the smooth integration into existing workflows.

Evolution of the direct laser engraving process – a faster, fully digital, much simpler alternative to the conventional exposure method has also encouraged the greater use of screen.

Direct laser engraving saves time because there is no need for the exposure, washing and drying processes, necessary for imaging the screens in the conventional way. Direct laser engraving systems achieve excellent quality, with maximum resolution of 2540 dpi, with very high contrast and precision in reproduction of positive and negative structures.

The availability of the pure non-woven nickel screen format to a wider market, and the introduction of a galvanic screen produced via an electro-forming process that requires no wire mesh base are interesting developments.

It is clear that the role of screen printing in the label sector will continue to develop and it has already carved out significant niches.

Chapter 7

The gravure printing process

Intaglio gravure printing originated with the goldsmith engravers in about 1446.
The images were hand engraved onto copper, gold and silver and the recessed image was filled
with a black ink or enamel known as Niello and then pressed onto paper.

These early prints were used by the goldsmiths to display the range of engravings available to the customer. The goldsmiths not only engraved their products but also developed an etching method useing nitric acid.

In about 1640 a German engraver by the name of Von Seigen employed a new method called mezzotint and this was used to reproduce paintings in black and white and also in color. Engravers mastered the art of varying the depth of the engraving or etching which allowed differing shades of color to be achieved-exactly the same principle as today's gravure process.

The intaglio process was further developed in the early seventeenth century when intaglio printing, also known as gravure printing, began to use metal plates which carried the etched image which was then printed onto the substrate. The invention of photography led to the method of transferring a photo image onto a carbon tissue coated in a light-sensitive gelatine which allowed the etching of an image onto a steel, copper covered cylinder. This was the beginning of the modern rotogravure process. An example of early rotogravure was the introduction of newspaper supplements 1930s –1960s which carried rotogravure printed photographs. Prior to this newspapers had published very few photographs.

GRAVURE PRINTING IN LABELS

Gravure printing has usually been the chosen process most suitable for printing fine tones with a sharp dot formation giving a high quality result and best suited to the long run markets. The self–adhesive label market tends to service the short to medium run markets and the gravure process has never been considered by the label printer as the most suitable.

However the use of gravure in the label industry has been increasing, particularly as the facility to produce highly reflective metallic inks and specialised coatings by the gravure process means that gravure units are now used as individual units on combination label presses. Gravure label presses are now used for both wet glue and self-adhesive label manufacture, particularly for the long run paper and filmic label market.

GRAVURE PRESS CONFIGURATIONS

Gravure press configurations fall into three categories: sheet-fed presses, dedicated in-line presses and single units used in combination with other printing processes. Every gravure printing unit comprises of two cylinders, a printing cylinder which carries the image, and a rubber covered impression cylinder. The web or sheet travels through these two cylinders and considerable pressure is applied between the two cylinders causing the ink in the recessed image cells to be transferred onto the substrate.

SHEET-FED

Sheet-fed gravure is rarely used in self-adhesive label printing. Historically it was used for the production of wet glue labels but has now largely died out. The configuration of the sheet-fed gravure press is very similar to the litho press with a feeder and sheet delivery system. However the gravure press will usually have only one printing unit and not multiple print heads and therefore will print only one color at a time. Because the gravure ink is a solvented system the sheet-fed press will have an extended sheet delivery system into which a hot air drying system is located, this will also include an extraction system to ensure that the solvent vapor is removed.

IN-LINE GRAVURE PRESSES

All the gravure press configurations used for self-adhesive label manufacture are in-line This means that there is a one printing unit per color and there are no real limitations on the number of printing heads used

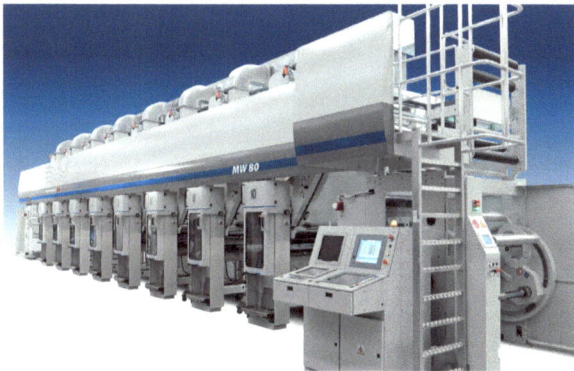

Figure 7.1 - Modern gravure printing press. Source: Bobst Rotomec

on an in-line press, **(See Figure 7.1).** One of the major differences between the in-line gravure press and the in-line litho and flexo presses is the position of the drying system. The web path of the litho and flexo press travels downwards after leaving each printing head. This takes the web through the infra-red or UV drying units which are positioned 'below' the print unit. The drying heads on the gravure press are positioned 'above' each print head which means that the web will travel upwards after leaving each printing head.

SINGLE GRAVURE UNITS

This type of gravure unit is usually incorporated into a dedicated in-line litho or flexo press or a multi-process platform press. The units, which are in a fixed position, are located in the press to give the best position to achieve the optimum printed result for highly reflective metallic inks and special coatings. In **Figure 7.2** you can see the configuration of the gravure unit of the type used in a multi-process combination press with the web, after printing, traveling upwards into the drying head and then passing downwards over a chilled roller and then into the next print unit for printing the next color.

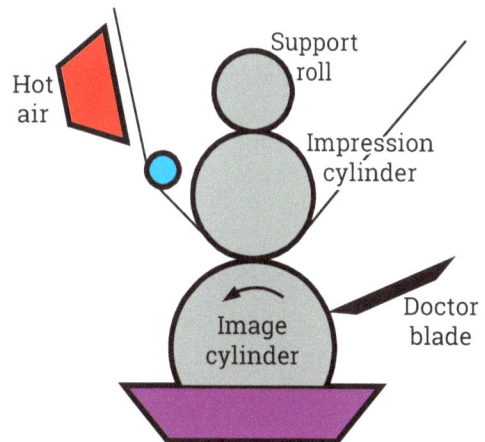

Figure 7.2 - Principles of gravure process

PRINCIPLES OF PROCESS

Unlike letterpress and flexo printing which are relief processes, the gravure process prints directly onto the substrate. The image is transferred directly to the substrate from the recessed image which is etched onto the gravure cylinder **(See Figure 7.3).** The gravure cylinder revolves in a liquid ink with the ink

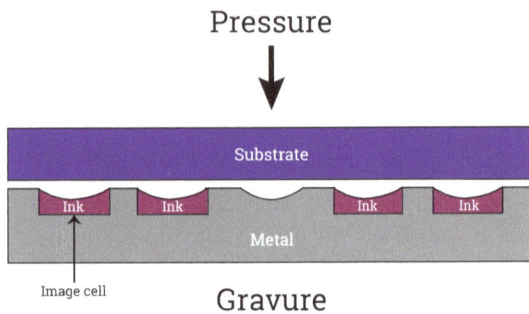

Figure 7.3 - Principles of gravure process

being applied to the whole of the imaged print cylinder i.e. there are no inking rollers, ink duct or anilox rollers used in the gravure unit.

A doctor blade assembly holds the doctor blade, which is in constant contact with the rotating print cylinder. The 'doctoring' action removes the surplus ink from the non-image areas leaving fresh ink in the cell cavities, The doctor blade is positioned close to the point where the two cylinders are in contact with the substrate which is sandwiched between these cylinders. **(See Figure 7.4).** This is called the 'nip' area and is the point at which the ink from the recessed cells is transferred to the substrate.

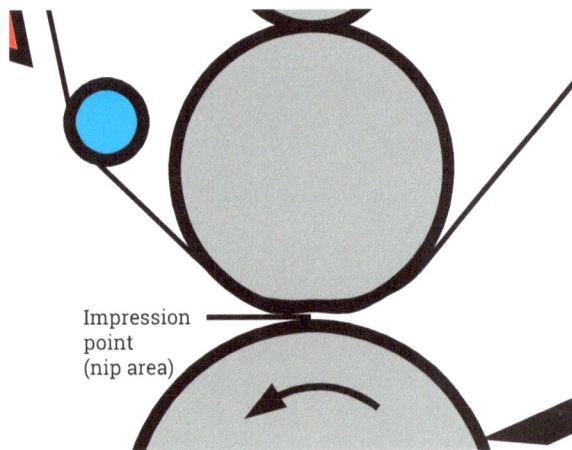

Figure 7.4 - Close up of point of print

The impression roller applies the downward force to the print cylinder. This impression roller is also in contact with an additional roller called the 'support roller' which gives additional support to the impression roller. **(See Figure 7.2).** The surface of the impression roller is rubber covered and the 'shore hardness' of the rubber can be varied to increase or decrease the dwell time ensuring that there is maximum dwell at the point of contact between the substrate, impression cylinder and the image cylinder. The type of shore hardness will vary with the type of substrate being printed.

After the printing sequence the web then travels upwards into the dryer system, and it is important that the liquid ink is completely dry before the next printing sequence.

The transfer of the ink from the cell to the substrate is a capillary action. It is important that the ink is fully transferred from the cell and there is no residual dried ink left in the base of the cell. One of the problems with the gravure process is a printing defect called 'dot skipping' and results in halftone dots being missed on the printed image. This is a result of a poor transfer of ink from the image cells to the substrate. The ink transfer can also be influenced by the absorbency of the substrate being printed.

Some gravure presses are equipped with an electrostatic assist system which operates by creating a positive and negative static charge between the ink and impression roller and the image cylinder which can assist in giving a smoother and more complete transfer of the ink from cell to substrate. Electro assist generates an electric field in the 'nip' area.

The chemical composition of the gravure ink can also influence the effectiveness of the electro-static assist system.

Gravure printing can also be affected by the type of substrate being printed; a smooth surface allows a better contact between the image cells and the substrate surface giving a better ink transfer and less dot skipping. A rougher and less smooth substrate can have an adverse effect on the ink transfer resulting in missing dots and poor print. Electro-static assist systems do offer the gravure printer some excellent benefits including improvements in print quality on less expensive stock, faster press speeds and less substrate waste.

IMAGING THE GRAVURE CYLINDER

There are three methods of imaging the gravure cylinder,

1. Traditional method photographically using carbon tissue and acid etch called chemical etching
2. Mechanical engraving
3. Laser engraving

The gravure cylinder comprises of a steel base, then a layer of copper and finally a polished chrome surface which is applied after the imaging process.

The image area on the cylinder is produced by engraving a cell formation that differs in depth and size. This variation in the cell structure controls the amount of ink and therefore the strength of color being printed, because the 'cell size' or 'dot size' can be engraved to a very small specification, subtle variations in dot/tonal detail and color strength can be easily produced.

The use of chemical etching for gravure printing within the label industry is very limited and this chapter will focus on the two main engraving systems used today i.e. mechanical and laser.

MECHANICALLY CUT CYLINDERS

The mechanical engraving of gravure cylinders started around the 1960s and is called a Klischograph system. This method of engraving requires the image cylinder to be revolving while a diamond tipped stylus, which is oscillating at 4000 times a second, pierces the surface of the image cylinder. Every time the stylus penetrates the copper it removes a chip from the surface creating a cell shaped indentation. The deeper the stylus penetrates, the larger the cell, the lighter the penetration the smaller the cell. The image to be engraved is scanned and the data is digitized. The digital information controls the engraving section of the machine. The depth of penetration of the stylus is controlled by the digital information and follows the tones and densities of the original artwork.

The cell structure of the mechanically engraved cylinder is governed by the shape of the stylus head and is a four sided pyramidal shape, the same as the cell formation on the mechanically engraved anilox rollers used in flexo printing. **(See Figure 7.5).**

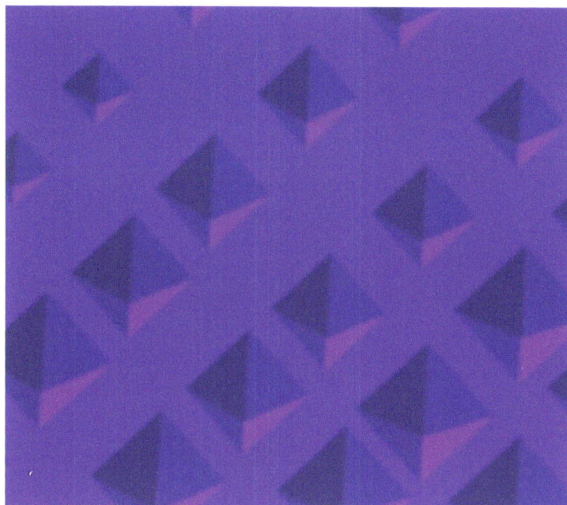

Figure 7.5 - Typical gravure cell formation

LASER ENGRAVED GRAVURE CYLINDERS

Direct laser engraving for gravure cylinder imaging is now widely used in the commercial, packaging and label markets.

Direct laser engraved cylinders are imaged using a laser engraving unit that is controlled by a digital file. The file contains all the data required to reproduce the exact tonal values and detail that are required for the correct image reproduction. Laser engraving removes the problem of image inconsistency when a duplicate cylinder is required. The cell shape of the laser engraved cylinder differs from the mechanical system. Whereas the mechanical cell is a pyramid shape, the laser is a scoop shape. The depth and width of the cell determines the volume of ink delivered **(see Figure 7.6).**

Gravure imaged cylinders or sleeves are manufactured with a copper outer surface on a steel base, If the cylinder is to be imaged with a mechanical engraving system then the copper layer is sufficient, however if the cylinder is to be direct laser engraved then an additional zinc layer is required.

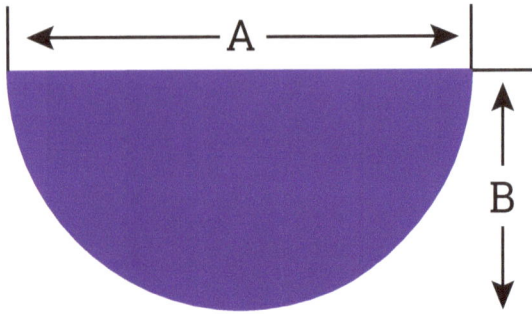

Calculation of cell volume

$$\frac{\text{Cell depth (B)}}{\text{Cell opening (A)}} \times 100\%$$

Figure 7.6 - Cell shape and ink volume

The laser engraves directly onto the zinc surface of the cylinder and is considerably faster than mechanical engraving, producing a cell formation at 70,000 cells per second. The facility to produce varying cell size and also cell shapes means that the cells can be tailored to produce very high resolution tonal values, particularly for vignettes with no 'saw edge' effect on small text and a much improved ink transfer.

After the engraving process is completed it is necessary to chrome the cylinder surface. This process of chroming gives the image cylinder a much harder surface, extending the life of the cylinder image and improving the doctoring process by ensuring that all the surplus ink is removed from the surface of the non-image area.

THE DOCTOR BLADE

As previously explained the image cylinder rotates in the ink duct coating the whole surface of the image cylinder with liquid ink including the image cells.

The ink on the non-image area is 'doctored' clean (wiped) by the doctor blade which is positioned over the cylinder **(See Figure 7.7)** at a point just prior to the nip area. The positioning of the doctor blade assembly is important. Because gravure inks are a solvented system and dry very rapidly, it is necessary to keep the distance between the doctoring process and the print 'nip' to a minimum, ensuring that the ink remains as open and liquid as possible.

The gravure doctor blade unit is far more sophisticated than the type used in the flexo system. The angle of attack to the cylinder can be adjusted from a 90 degree angle to a much sharper 45 degree angle and this facility allows the printer to achieve the optimum position for a clean wipe and the best possible printed result. The thickness of the doctor blade can also be varied to assist in achieving the best doctoring result.

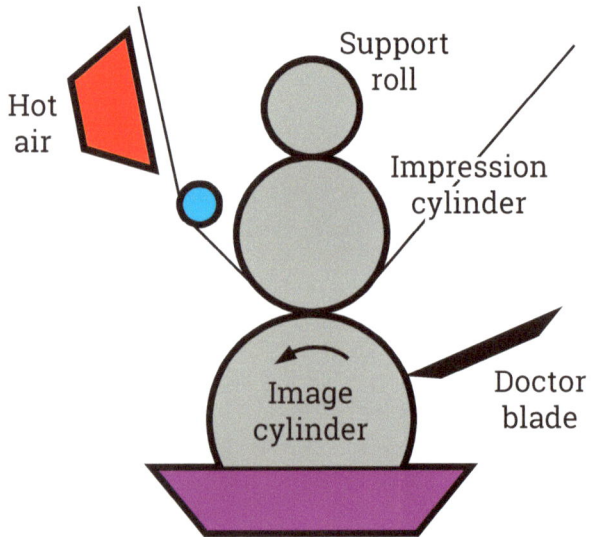

Figure 7.7 - Location of docter blade on gravure unit

By varying the thickness of the doctor blade the printer can also control the volume of ink left in the cell. A thinner blade removes slightly more ink from the cell and albeit this is a very small amount it can slightly reduce the density of the color being printed.

The steel doctor blade is supported by a backing blade to give better stability. Traditionally the doctor blade would be removed and re-sharpened by the printer using an oiled stone, and this would be done at intervals throughout the print run. However the modern doctor blade is now a replaceable item which removes the need for re-sharpening by hand.

GRAVURE INKS AND DRYING SYSTEMS
The majority of inks used in the gravure process are solvent based, however there are some gravure printed self-adhesive labels that are printed using UV gravure inks.

Solvent based inks require infra-red and/or hot air drying to drive off the solvent content. Because the inks are liquid and very fast drying it is necessary for the ink to be constantly circulated through the ink weir, ink trough and the ink pumping system. Maintaining the correct ink viscosity is critical and the viscosity readings can be made manually or automatically. Adjustments can be made to the speed of the ink drying by the introduction of a faster or slower drying solvent, however this adjustment must be done under tight control as it is important that the correct balance of the solvents which make up the ink formula is maintained. Gravure inks can be formulated using different types of solvent systems and the printer must ensure that any solvent used is of the correct system.

The drying system on the gravure label press differs from the systems used on litho, flexo and screen presses. The drying unit is positioned above the print unit meaning that after the print sequence the web travels upwards into the dryer. **(See Figure 7.8).**

IDENTIFYING THE GRAVURE PROCESS
The gravure process can be easily identified by the saw tooth or serrated edge visible on the printed text and on the edges of solid color areas.

This saw edge is a characteristic of the process and is created by the outer edge section of the individual cells which make up the image. The screen process has a similar characteristic. Gravure and screen imaging cannot produce 'half a screen' and therefore cannot produce a straight edge.

However, this is not the case with laser imaged cylinders as the laser can engrave half a cell,

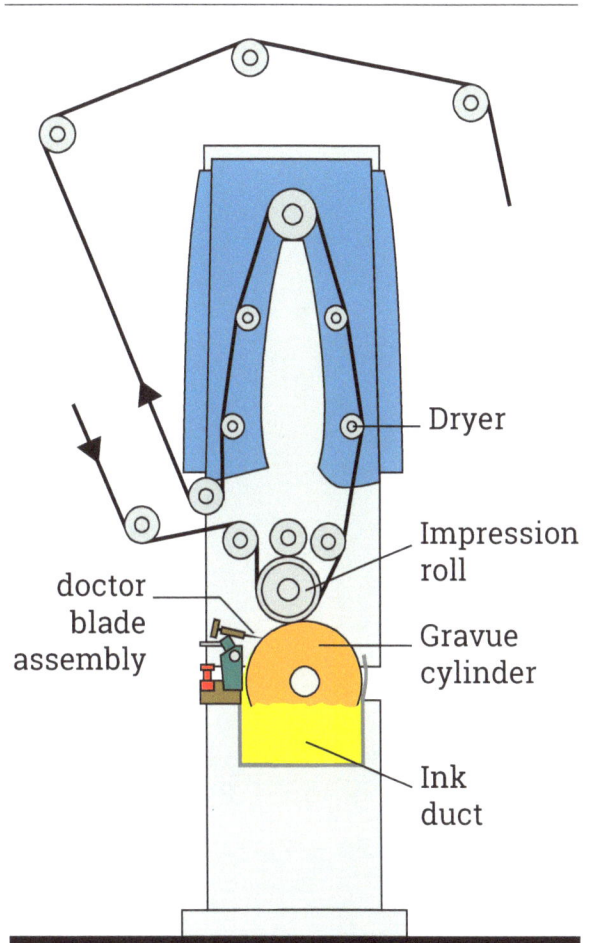

Figure 7.8 - Gravure unit showing drying head.

producing a straight edge. This control over the engraving process illustrates the level of print quality that can be produced by the gravure process.

Another method of identifying the gravure process is the reticulation that can occur within the solid area of the print.

Reticulation can be created by inks that are either too thin in viscosity or the solvent balance is incorrect.

ADVANTAGES OF THE GRAVURE PROCESS
- Sheet or web-fed application
- Excellent print quality - excellent sharpness

and reproduction
- High color consistency
- Precise inking with no color variation throughout the print run
- Consistent and substantial ink film weight
- High speed printing
- Wet on dry - excellent half-tone reproduction
- Long cylinder life - making repeat runs very economical
- Suitable for food packaging
- Most suitable process for printing of metallic inks
- Gravure cylinder sleeve option
- Laser engraving high quality - 70,000 cells per second variable screen
- Exact cylinder/image replication

DISADVANTAGES OF THE GRAVURE PROCESS
- High preparatory and origination costs, but due to improved technology these have reduced
- High capital cost of press
- Most gravure ink systems solvent based (environmental issues)

GRAVURE PRINTED APPLICATIONS
- Self-adhesive labels. filmic – paper – metallic
- Filmic materials
- Flexible packaging and shrink sleeves
- Magazines/catalogues/books
- Shrink sleeves
- Wallpaper
- Security printing eg stamps, banknotes, bonds,
- Fine art (sheet-fed)
- Food packaging
- Wrapping paper
- Furniture laminates
- Paneling
- Greeting cards

GRAVURE – SUMMARY
Gravure has traditionally been used as a long run process for wet-glue labels.

More recently it has made inroads into narrow and mid-web presses for self-adhesive labels where the ability of gravure to apply varnishes , lacquers and top coatings offers particular advantages.

Gravure units are typically added as part of a combination press configuration.

DEVELOPMENTS IN METALLICS
Gravure is an excellent process for applying metallic inks and finishes with a high level of lustre.

Developments exist where vacuum metallised particle inks are reverse printed by the gravure process onto a clear substrate, such as a PP.

This method gives a very bright and highly reflective appearance resembling that of foil blocking. The metallised particles are easily suspended in the solvent medium used in the gravure process and align smoothly and easily to the clear label substrate.

PLATE SYSTEMS
Photopolymer plate systems for gravure have been developed as well as computer-to-cylinder systems.

An 'etched' polymer plate can be produced with indentations of varying depth which 'hold' the ink. Depending on the quality, the resulting print may look similar to, or the same as those produced with the traditional photogravure process using etched cylinders.

More recently a direct digital laser etching process has been introduced, which will eventually reduce the cost of imaging gravure cylinders.

Chapter 8

The combination press

Technical developments within the label industry have given printers and converters the opportunity to utilise all the existing printing processes within a 'multi-platform press'. The combination or platform printing press is a highly engineered, sophisticated machine **(See Figure 8.1)** which uses a system of process cassettes which are interchangeable between each base unit.

This flexibility facilitates the printing, embellishing and converting of self-adhesive labels in a single pass. Being able to apply a variety of processes onto the same label allows the designer to exploit the distinct advantages of each process in order to create innovative graphics and effects.

The embossing, hot foil stamping and lamination processes are also in a cassette format and this adds further opportunities to increase the added value content of the label. This flexibility linked to in-line profile die-cutting, sheeting and slitting offers some major advantages to the label printer and converter.

Figure 8.1 - Modern combination press. *Source: Omet*

COMBINATION PRINTING - KEY DRIVERS

Innovative packaging design is a critical factor in the success of any brand.

Indeed product decoration plays a vital role in establishing a competitive edge to a product and in encouraging purchases at the point of sale.

The development of combination printing has provided designers with new opportunities to create the unique graphics that they are seeking and it has been this driver that has stimulated the substantial growth in the use of multi-process graphics.

Combination printing is now widely used to produce self-adhesive labels in a variety of market sectors including;

- Wines and spirits
- Security
- Beverages
- Cosmetics
- Personal care products
- Pharmaceuticals
- Promotional graphics

WHY COMBINATION PRINTING?

Why is there a need to use different print processes on a single production run?

Each of the various print technologies has its own

advantages and disadvantages. Combination printing focuses on the advantages offered by each process allowing the printer/designer to utilise these to increase the visual appeal of the label.

The advantage of each process have been clearly identified in each of the relevant chapters, but a typical example would be the difference in print characteristics between the offset litho process and the screen process. Litho will produce very fine tonal detail, but cannot deliver high coating weights of ink, whereas the screen process will produce very high coating weights of ink, but cannot produce the very fine tonal detail.

The graphic processes that can be used in any combination are:

- Offset litho
- Rotary silk screen
- UV and water-based flexo
- Letterpress
- Gravure
- Digital
- Foiling
- Laminating
- Embossing
- Varnishing

THE COMBINATION PRESS

There are two types of combination label press;

1. Different print technologies with the print heads in fixed positions on the press,
2. The open platform type with the facility to exchange different processes in the same platform unit.

The fixed position press does not have the flexibility to move a process from one unit to another. The position of each process is determined when the press is manufactured.

A typical configuration for this type of press would be 1 x screen and 6 x flexo, with screen in unit 1 and flexo in units 2-3-4-5-6. This print sequence would allow the printing of a heavy coating of white using the screen process, followed by line, text and tones printed flexo. This sequence would be typical for the production of a clear filmic self-adhesive label.

This chapter will look in detail at the second type of

combination press, the fully interchangeable system either cassette system or a sleeved changeable system.

The modern combination press is an in-line configuration and can be equipped with all the facilities and ancillary items highlighted in chapter 2 (Press Configurations). It is a highly specified piece of equipment and typical press 'control' system would include:

- Central press control unit with touchscreen
- Central UV-system control
- Press error diagnostic
- Automatic Mark-to-Print register control system
- Automatic electronic system – length and side register, including self-adjusting features and history storage
- Advanced PMC for integration of video web inspection system and monitor for remote computerised ink zone control
- Job management package
- Register log program for customer quality reports
- External back-up function, inclusive of software and transferable memory
- On-line service function

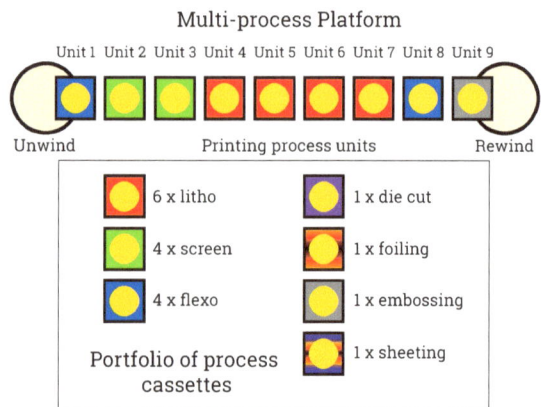

Figure 8.2 - Combination press with multiple print processes

COMBINATION PRINTING PRESS- PROCESS FLEXIBILITY

The Illustrations in **Figure 8.2** and **8.3** show the flexibility offered by a combination press. **Figure 8.2** shows the platform with a print unit sequence of 1 x flexo, 2 x screen, 4 x litho and 1 x flexo and 1x embossing unit.

Figure 8.3 shows the same platform press but with a print sequence of 4 x litho, 1 x screen, 2 x litho, 1 x flexo and 1 x embossing unit.

As the modern combination press uses digital driven servo motors for all the press functions, very accurate print to print, print to embellishment and conversion registration is maintained regardless of the

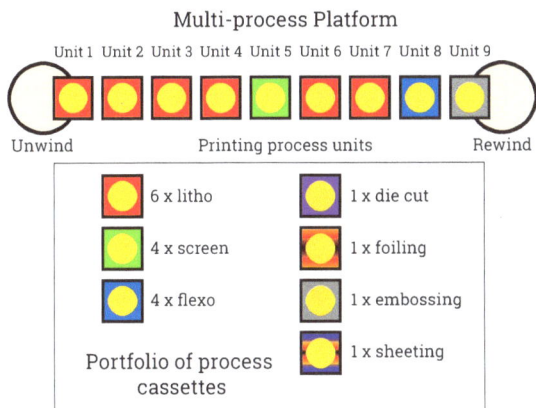

Multi-process Platform

Unit 1 Unit 2 Unit 3 Unit 4 Unit 5 Unit 6 Unit 7 Unit 8 Unit 9

Unwind   Printing process units   Rewind

6 x litho		1 x die cut	
4 x screen		1 x foiling	
4 x flexo		1 x embossing	
Portfolio of process cassettes		1 x sheeting	

Figure 8.3 - Same combination press but with process sequence changed

type of substrate in use. Early combination presses experienced difficulties with process registration and ink compatibility and development work was also required to establish the correct repro distortion factors between each process to ensure that the print length of each different process was exact.

GRAVURE-INTAGLIO PRINTING ON THE COMBINATION PRESS

Over the last 10 years there has been an increase in the use of gravure printing units fitted to combination presses. These units are not interchangeable with the other processes units and the position of the gravure

unit has to be established at the manufacturing stage and remains in a fixed position on the press. It is not uncommon for combination presses to be equipped with more than one unit with some combination presses running with three gravure units. One of the major benefits of the process is the outstanding results which can be achieved by gravure when printing silvers, golds and other metallic colors, using solvent based inks. This gives the printer a much less expensive option of achieving metallic embellishments than hot/cold foiling.

There is also an increasing trend in the use of what is known as 'intaglio' gravure. The difference between a gravure print and an intaglio print is the method of imaging the image cylinder. The gravure image can be acid etched, mechanically engraved or laser engraved and is made up of a series of cells. The intaglio image is hand engraved using an engraving tool and the image is created by the channels cut with the engraving tool. The ink is transferred from the engraved channels onto the substrate and when dried leaves a slightly raised image.

The intaglio process offers a printed effect that is difficult to copy and is used extensively as an added security measure for the stamp and banknote industry.

THE CASSETTE SYSTEM

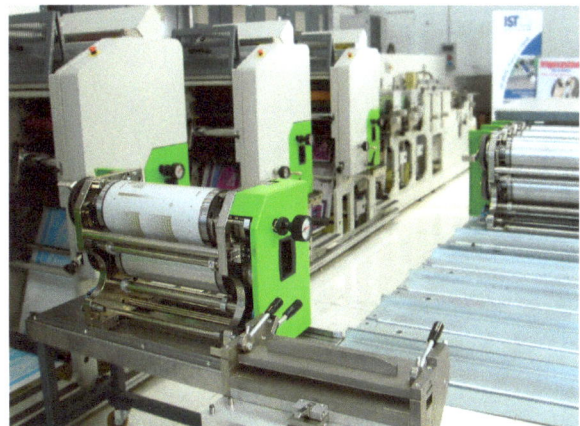

Figure 8.4 - On-press cassette change. *Source: Gidue*

The cassette system is a quick and uncomplicated method of changing the type of process used on an individual base unit. **(See Figures 8.4).** All the pneumatic, electrical and electronic connections and links can be easily disconnected and reconnected, with the average cassette changeover time taking about 2-3 minutes. The down side to the cassette system is the need to have a range of cassettes which cover both the number of units required for each printing process and in the case of the litho process, a number of different fixed, plate and blanket cylinder sizes to cover the range of print repeat lengths required.

THE SLEEVED - CHANGEABLE SYSTEM

The use of sleeves in the flexo process has been commonplace for some time. The system has now been developed for narrow-web offset presses by replacing both the fixed position blanket and plate cylinder. The sleeves are simply inserted onto a pneumatic shaft. **(See Figure 8.5).**

Figure 8.5 - Litho cylinder change using sleeve system. Source: Muller Martini

This sleeve development, plus the introduction of stepped motor technology and digital driven control systems, has given the press the ability to make easy changes to the print length without the need for a fixed diameter cassette. This system gives the label printer a more cost effective method of facilitating differing repeat lengths and job changeover can be carried out

within minutes. The sleeve system eliminates the need to have dedicated process interchangeable cassettes.

INKS AND DRYING SYSTEMS FOR COMBINATION PRINTING

The ink rheology and the drying/curing systems used for combination printing are the same as the inks and drying systems described in each of the chapters relating to each individual process.

Inks from each of the printing processes are now compatable with few problems. Printing will be a wet on dry application. Every process except the gravure process uses the same drying curing system i.e UV or hot air. The gravure unit however will have its own hot air delivery and extraction system.

IDENTIFYING COMBINATION PRINT

Identifying labels that have been printed in combination can be difficult because it is necessary to identify each individual process and then work out the print sequence (ie the order in which the label has been printed).

INVESTING IN COMBINATION PRINTING - POINTS FOR CONSIDERATION

Careful consideration has to be given to a number of important areas before taking the decision to invest in a combination press. It is important that the relevant departments that make up the company structure (marketing, sales, pre-press and the production teams) are all involved in the project at the early stages.

A portfolio of existing and 'potential' customers needs to be assembled and a work content analysis carried out, based on the feasibility of the work being produced on a combination press and the potential sales increase that could be generated as a result. The modern label press now has the capability to produce a number of decorative applications; self-adhesive, IML (in-mold labels) and wet glue labels, cartons and shrink sleeves and all these market opportunities must be given due consideration.

A skills assessment is important as the introduction of a combination press will put extra demands on the skill base of the company.

The more experience the company has in the different printing processes the easier the learning

	Litho	Letterpress	Gravure	Flexo	Silk Screen	Digital
Ink Weights	3	5	8	8	10	2
Print Quality/ Tonal Value	10	6	9	8	4	10
Ink Opacity	2	6	6	8	10	2
Preparatory Costs	2	4	10	4	8	2
Running Speeds	9	6	10	9	2	2
Product Resistance	8	8	8	8	10	2
Plate Life	4	8	10	6	6	-

Figure 8.6 - Guide to relative strengths and weaknesses of each print process

curve and therefore a training program should be developed that overcomes any skill shortfall.

COMBINATION PRINTING - COST CONSIDERATIONS

The cost of introducing a combination press can be considerable, not just the initial cost of the press, but also the investment required for the press support equipment and the potential increase in the overhead costs of each department.

The cost implications fall into two categories;

1. The costs of the press, the press ancillary equipment, any additional print process equipment, any pre-press upgrading, any upgrading of power on press should remember that this can leave the company vulnerable should there be any press supply and increased press maintenance.

2. The training costs incurred to achieve the skill levels required for combination printing. Higher skill levels are required for the sales and marketing staff to understand the benefits and potential of the combination press. There is greater complexity within the pre-press/repro function and higher skill levels related to the press, press support and the effect on press maintenance.

LIMITATION ON CAPACITY WITH A SINGLE COMBINATION PRESS.

The ability to produce multiple graphics and embellishments on a wide range of substrates in a one pass operation is unique compared to other packaging decoration systems.

Any company anticipating the installation of a single combination press should remember that this can leave the company vulnerable should there be any press breakdowns.

PRINT PROCESS COMPARISON

The chart **(See Figure 8.6)** is a simple way to identify the strengths and weaknesses of the printing processes that can be used in combination printing. It shows the comparison between processes in seven areas and scores each process on a 1- 10 basis with 10 being the highest rating and 1 being the lowest.

SOME DIFFERENT PRESS CONFIGURATIONS

The following diagrams are press configurations that can be used in combination printing. Each diagram has a brief description of the application and the appropriate web path and the print processes used.

Figure 8.7 shows the combination press with a portfolio of nine cassettes: 2 x screen, 1 x flexo for reverse printing and 6 x flexo units. The press is fitted with an overhead system onto which is located a lamination unit which would run a self-adhesive laminate.

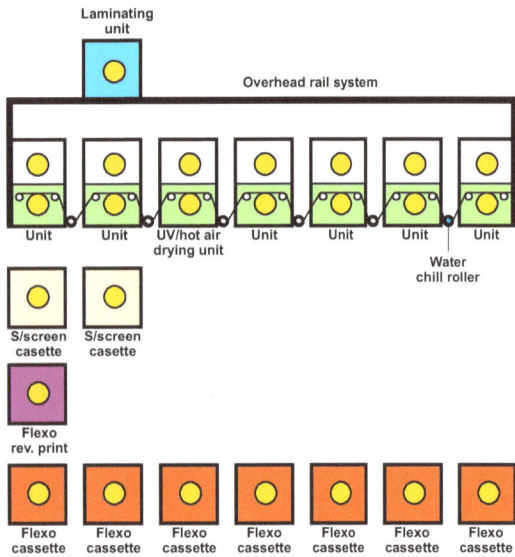

Figure 8.7 - 7 color combination press with overhead rail system

Figure 8.8 shows the same press with two screen units located onto the overhead, but with the addition of a drying unit. In both **Figure 8.7** and **Figure 8.8** the press is using a single unwind unit and the web path is re-positioned to accomodate the print sequence.

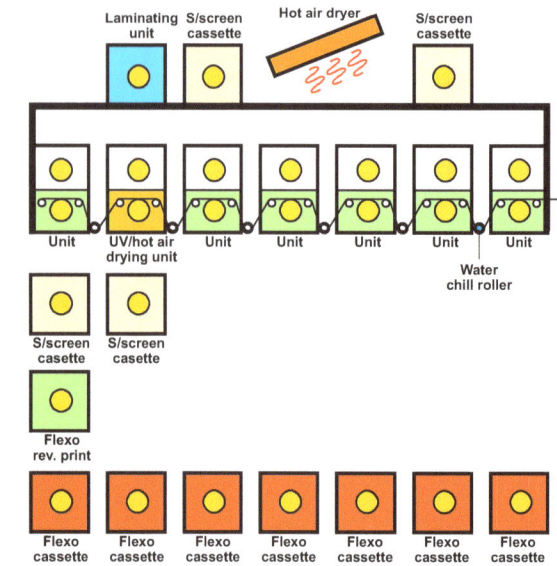

Figure 8.8 - Combination Press with Overhead Rail System

PRINTING ON THE LAMINATE

Figure 8.9 shows a press with two unwinds, one with a self-adhesive substrate and one with a film laminate. The press is fitted with an overhead system onto which two screen printing units are located. The press

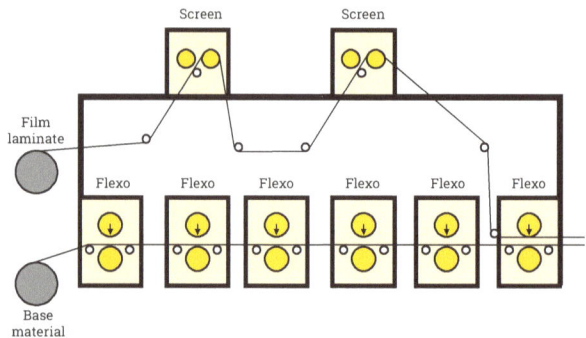

Figure 8.9 - Printing onto a clear overlaminate

is a dedicated six color press using the flexo process. The overhead screen units are printing two colors on

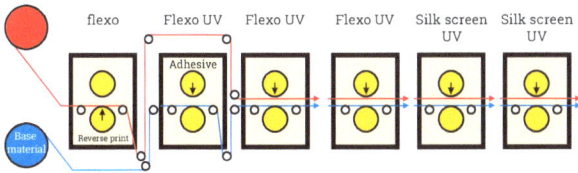

Figure 8.10 - Combination print with reverse printed laminate

the face of the laminate and the press is printing four colors in units 1-2-3-4- onto the self-adhesive substrate. Unit 5 is applying an adhesive to the self-adhesive substrate and unit six is compressing the

two substrates to form the laminated label. **Figure 8.10** shows a 6 color combination press with two unwind units. Unit one is printing on the reverse of a clear laminate substrate, using a high reflective metallic ink. The self-adhesive substrate enters unit 2 from the underside, by-passing unit 1. The laminate web is directed over the top of unit 2, the face of the self-adhesive substrate is coated with adhesive in unit 2 and the two substrates are compressed together in the nip roller positioned between units 1 and 2. Units 3-4-5-6 are printing both screen and flexo onto the surface of the laminate. The printed image on the reverse side of the laminate is viewed from the front of the label, giving the metallic ink a very highly reflective appearance similar to hot foil stamping.

Index

www.ingramcontent.com/pod-product-compliance
Lightning Source LLC
Chambersburg PA
CBHW041722210326
41598CB00007B/741